FLORA OF THE GUIANAS

Edited by

M.J. JANSEN-JACOBS

Series A: Phanerogams

Fascicle 22

T0141381

27. PHYTOLACCACEAE
(+ remarks 28. ACHATOCARPACEAE)
29. NYCTAGINACEAE
30. AIZOACEAE
32. CHENOPODIACEAE
33. AMARANTHACEAE
34. PORTULACACEAE
35. BASELLACEAE
36. MOLLUGINACEAE
37. CARYOPHYLLACEAE
(R.A. DeFilipps & S.L. Maina)

54. SARRACENIACEAE
(A.S.J. van Proosdij)

55. DROSERACEAE
(R. Duno de Stefano)

including
Wood and Timber
(J. Koek-Noorman & P. Détienne)

2003
Royal Botanic Gardens, Kew

Contents

© The Trustees of The Royal Botanic Gardens, Kew.
ISBN 1 84246 069 2

Printed and Bound by The Cromwell Press Ltd.

KEY TO THE FAMILIES OF THE CARYOPHYLLALES
by
ROBERT A. DEFILIPPS[1]

1 Stems succulent, armed; leaves usually absent . . . .
. . . . 31. CACTACEAE (in fasc. 18 Flora of the Guianas, 1997)
Stems not succulent, usually unarmed (if armed, then leaves present); leaves
present 2

2 Carpels 1-several, distinct, ovule 1 per carpel, or carpels 2 or more, united in
a compound ovary with as many locules and ovules as carpels 3
Carpels always 2 or more, united in a compound ovary, ovules more than
1 per carpel 4

3 Leaves alternate; inflorescence not subtended by a involucre; tepals free, not
petaloid . . . . 27. PHYTOLACCACEAE
Leaves often opposite; inflorescence often subtended by a conspicuous
involucre; perianth connate in a petaloid tube simulating a corolla . . . . . .
. . . . 29. NYCTAGINACEAE

4 Leaves succulent; ovary superior or inferior . . . . 30. AIZOACEAE
Leaves usually not succulent; ovary superior 5

5 Perianth with only tepals, not petaloid 6
Perianth with sepals and petals, or sepaloid bracteoles and tepals 7

6 Perianth greenish, herbaceous; filaments free . . . 32. CHENOPODIACEAE
Perianth scarious, dry; filaments connate below in a tube . . . .
. . . . 33. AMARANTHACEAE

7 Sepals or sepaloid bracteoles 2 8
Sepals 4 or 5 9

8 Herbs or subshrubs; ovules 2-several; fruit a capsule . . . .
. . . . 34. PORTULACACEAE
Vines; ovule 1; fruit indehiscent . . . . 35. BASELLACEAE

9 Leaves alternate, opposite or whorled; petals absent or inconspicuous; ovary
with 2-several locules . . . . 36. MOLLUGINACEAE
Leaves opposite; petals usually present, conspicuous; ovary with 1 locule . . .
. . . . 37. CARYOPHYLLACEAE

[1] National Museum of Natural History, Department of Systematic Biology - Botany, NHB 166, Smithsonian Institution, Washington, D.C. 20013-7012, U.S.A.

27. PHYTOLACCACEAE
(with remarks on **28. ACHATOCARPACEAE**)
by
ROBERT A. DEFILIPPS AND SHIRLEY L. MAINA[2]

Herbs, shrubs or lianas (vines), rarely trees. Leaves alternate, simple, entire, usually petiolate; petiole deeply sulcate adaxially; stipules present or absent. Inflorescence a terminal or pseudoaxillary raceme, spiciform raceme, spike or panicle; pedicels bracteate. Flowers bisexual or unisexual, regular; tepals 4 or 5, united below in a 4- or 5-lobed perianth, inconspicuous or petaloid; stamens 3-20, free or the filaments united at the base, filaments in 1 or 2 whorls, sometimes borne on a hypogynous disk, anthers 2-locular, dorsi- or basifixed, longitudinally dehiscent; ovary superior or half-inferior, 1- to 16-locular, free or partly to wholly connate carpels, ovule 1 per carpel, basal in single carpels, axile in syncarpous ovaries, campylotropous, style, if present, as many as carpels, usually free, stigma capitate, or sessile and penicillate. Fruit a berry, achene, capsule, samara or drupe; seeds 1-numerous, perisperm copious to absent, embryo curved.

Distribution: A mostly neotropical family of 17 genera and approximately 70-120 species; in the Guianas 6 genera and 9 species.

Note: ACHATOCARPACEAE is closely related to PHYTOLACCACEAE, differing from the latter by the absence of raphides and styloids, the morphology of the sieve element plastids, the compound, 1-ovulate ovaries and the absence of anomalous secondary growth (Bittrich, 1993). The genus *Achatocarpus* Triana ranges from Texas and Mexico south to Argentina, and has an unconfirmed report from the Guianas. No Guianan specimens have been received for study. The South American *Achatocarpus pubescens* C.H. Wright, with racemes to 4.5 cm long and 2 stigmas, superficially resembles *Trichostigma octandrum* (L.) H. Walt. (PHYTOLACCACEAE) which has racemes to 13 cm long and 1 stigma.

[2] National Museum of Natural History, Department of Systematic Biology - Botany, NHB 166, Smithsonian Institution, Washington, D.C. 20013-7012, U.S.A.

LITERATURE

Bittrich, V. 1993. Achatocarpaceae. In K. Kubitzki, The Families and Genera of Vascular Plants 2: 35-36.

Brown, G.K. & G.S. Varadarajan. 1985. Studies in Caryophyllales I: Reevaluation of classification of Phytolaccaceae s.l. Syst. Bot. 10: 49-63.

Correll, D.S. & H.B. Correll. 1982. Flora of the Bahama Archipelago, Phytolaccaceae. pp. 499-505.

Grenand, P. *et al.* 1987. Pharmacopées traditionnelles en Guyane, Phytolaccaceae. pp. 353-355.

Guaglianone, E.R. 1996. Phytolacca rivinoides (Phytolaccaceae), su presencia en la Argentina. Darwiniana 34: 399-401.

Jarvis, C.E. *et al.* 1993. A List of Linnaean Generic Names and their Types. Regnum Veg. 127. 100 pp.

Lachman-White, D.A. *et al.* 1987. A Guide to the Medicinal Plants of Coastal Guyana. Commonwealth Science Council Technical Publication Series No. 225. 350 pp.

Lemée, A.M.V. 1955. Flore de la Guyane Française, Phytolaccacées. 1: 575-579.

Nowicke, J.W. 1968. Palynotaxonomic study of the Phytolaccaceae. Ann. Missouri Bot. Gard. 55: 294-363.

Ostendorf, F.W. 1962. Nuttige Planten en Sierplanten in Suriname, Phytolaccaceae. Bull. Landbouwproefstat. Suriname 79: 27-28.

Raeder, K. 1961. Phytolaccaceae. In R.E. Woodson & R.W. Schery, Flora of Panama 4: 408-421 = Ann. Missouri Bot. Gard. 48: 66-79.

Rogers, G.K. 1985. The genera of Phytolaccaceae in the southeastern United States. J. Arnold Arbor. 66: 1-37.

Rohwer, J.G. 1993. Phytolaccaceae. In K. Kubitzki, The Families and Genera of Vascular Plants 2: 506-515.

Santos, E. & B. Flaster. 1967. Fitolacáceas. In P.R. Reitz, Flora Ilustrada Catarinense. 37 pp.

Stoffers, A.L. 1957. Phytolaccaceae. In A.A. Pulle, Flora of Suriname 1(2): 209-217.

Walter, H. 1909. Phytolaccaceae. In H.G.A. Engler, Das Pflanzenreich IV. 83 (Heft 39): 1-154.

KEY TO THE GENERA

1 Ovary with 5-16 (1-ovulate) carpels; aggregated fruit with several seeds . . .
. . . . *3. Phytolacca*
  Ovary 1-locular, with 1 ovule; fruit with 1 seed 2

2 Stigmas 2, terminal . . . . *1. Microtea*
  Stigma 1, sometimes lateral 3

3 Flowers sessile or subsessile (inflorescence a spike-like raceme); ovary and fruit with 4-6 retrorse apical hook-like awns (crushed plant with garlic or onion odor) .... *2. Petiveria*
Flowers pedicellate (inflorescence a raceme); ovary without awns 4

4 Stem armed; fruit a samara. .... *5. Seguieria*
Stem unarmed; fruit a drupe 5

5 Annual, erect herbs or subshrubs; stamens 4; style elongate; tepals erect in fruit .... *4. Rivina*
Woody vines; stamens 8-12; style absent; tepals reflexed in fruit ....
.... *6. Trichostigma*

1. **MICROTEA** Sw., Prodr. 4, 53. 1788.
Type: M. debilis Sw.

Annual herbs, sprawling, sometimes slightly woody at base. Leaves petiolate or subpetiolate, stipules absent. Inflorescences terminal or extra-axillary spikes or spike-like racemes; bracts present; bracteoles 0 or 2. Flowers bisexual, sessile or shortly pedicellate; tepals 5, oblong, persistent in fruit; stamens 3-5, alternating with tepals, or 6-9 with 1-4 epitepalous ones, anthers dorsifixed, globose; ovary 2-carpellate, 1-loculed, globose, stipitate, ovule 1, basifixed, style absent or nearly so, stigmas 2, each sometimes 3-partite. Fruits small, thin-walled drupes or achenes, tuberculate, glochidiate, muricate, spiny or smooth; seed 1, lenticular, testa black, shining, crustaceous, embryo curved, albumen scanty.

Distribution: Approximately 10 species in tropical America (West Indies, Central and South America); 2 in the Guianas.

Note: The genus *Microtea* is sometimes included in the family CHENOPODIACEAE.

KEY TO THE SPECIES

1 Raceme rarely more than 4 cm long; stamens 5; fruit with pale, non-glochidiate spines .... *1. M. debilis*
Raceme up to 10 cm long; stamens 6-9; fruit with dark brown, glochidiate (tip recurved with 4 retrorse bristles) spines .... *2. M. maypurensis*

1. **Microtea debilis** Sw., Prodr. 53. 1788. Type: West Indies, St. Christopher, Swartz s.n. (BM, not seen). – Fig. 1 (G-H)

Narrowly taprooted herb. Stem ascending to prostrate, sharply angled, glabrous, to 50 cm long. Leaves elliptical or ovate (rhombic-ovate to oblanceolate), near base of plant to 8 x 3 cm, acute to acuminate, and apiculate at apex, cuneate at base, broadly decurrent on petiole, glabrous; petiole to 3 cm long. Inflorescence a 10- to 25-flowered raceme, 1.5-4(-5.5) cm long; bract seemingly only an enation, ovate, ca. 0.3 mm long, obtuse, green; pedicels 0.8-1.2 mm long; bracteole elliptical, ca. 0.8-1 mm long, scarious, often notched. Tepals lanceolate or ovate, 0.5-1.1 mm long, subacute or obtuse, white with green midvein; stamens 5, alternate with and reflexed between tepals, filaments ca. 0.5 mm long; style 1, stigmas 3-partite, linear, spreading. Achenes spreading from inflorescence-axis, greenish, globose, 1-1.5 mm in diam., echinulate in a honeycomb pattern; seed globose, ca.1.3 mm in diam., black, smooth.

Distribution: New World tropics (West Indies, Central and South America); in the Guianas in disturbed habitats such as lawns, pastures and sandy roadsides; 67 collections studied, all from the Guianas (GU: 25; SU: 24; FG: 18).

Selected specimens: Guyana: Pomeroon Distr., Moruka R., de la Cruz 4601 (US); Northwest Distr., Amakura R., de la Cruz 3548 (US). Suriname: Paramaribo, forest behind Gonggrijpstraat, Samuels 65 (MO); Saramacca Distr., Experimental Farm Coebiti, Everaarts 494 (U). French Guiana: Iles du Salut, Ile Royale, Feuillet 2181 (US); Saül, at Eaux Claires, Mori et al. 23295 (NY).

Uses: The plant is employed to remedy albumin in the urine and difficult urination (Ostendorf, 1962). In Guyana, as noted by Lachman-White et al. (1987), "The entire plant is used to make a tea which is used for palpitating heart conditions and to effect "cooling" of inflamed areas. The plant may be used with others, such as st. john's-bush (*Justicia secunda*) and inflammation-bush (*Vernonia cinerea*) as an abortifacient, it may be used as a lotion for ulcers, or, mixed with muniri-dan (*Siparuna guianensis*) and taken orally, in treating diabetes". Said to be used in Guyana as a cooling drink for the head (Warren s.n.). Formerly, in French Guiana an infusion made from the entire crushed plant was used in Creole medicine for a diuretic and hypotensive drink.

Vernacular names: Guyana: fat-of-the-earth. Suriname: eiwitblad, eiwit-wiwiri; fladdiot. French Guiana: alentou case, l'en tout case (Creole); eye bulu wiwiri (Boni); racine-pistache.

6

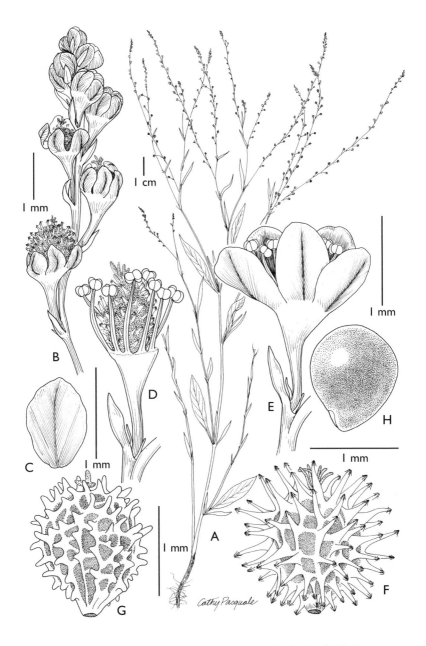

Fig. 1. *Microtea maypurensis* (Kunth) G. Don: A, habit; B, detail of infructescence; C, tepal; D, flower with perianth removed; E, flower; F, fruit. (B-E, Ro. Schomburgk ser. I, 116; F, Appun 2459). *Microtea debilis* Sw.: G, fruit; H, seed. Drawing by Cathy Pasquale.

2. **Microtea maypurensis** (Kunth) G. Don in Loudon, Hort. Brit. 98, n. 6423. 1830. – *Ancistrocarpus maypurensis* Kunth in Humb., Bonpl. & Kunth, Nov. Gen. Sp. ed. qu. 2: 186, t. 122. 1817. Type: not designated.
– Fig. 1 (A-F)

Erect to decumbent herb to 50 cm tall. Stem slender, glabrous, angular, suberect. Leaves lanceolate or linear-lanceolate, rarely elliptic-lanceolate, ca. 3.5 x 0.3-1 cm, acute at apex and base, glabrous; petiole ca. 1 cm long. Inflorescence a terminal, rarely axillary, pedunculate, elongate, suberect raceme to 10 cm long; pedicels ca. 1.5 mm long; bracts lanceolate, ca. 0.8 mm long, acute; bracteoles linear-subulate, ca. 0.3 mm long, inserted at base of pedicel. Tepals elliptical, ca. 1.5 mm long, acute, with green midvein; stamens 6-9, anthers ca. 0.2 mm wide; ovary ca. 0.5 mm, stigmas 3(5-6)-partite, ca. 0.2 mm. Achenes exceeding tepals, black, globose, ca. 1.5 mm, acute, echinulate in a honeycomb pattern, spinelets glochidiate, with recurved apex and 4 retrorse apical bristles.

Distribution: Tropical South America; in the Guianas found along riverbanks or in disturbed sites; 26 collections studied, 8 from the Guianas (GU: 5; SU: 3).

Selected specimens: Guyana: Rupununi Dist., Lethem, bank of Takutu R., Irwin 788 (US); Lower Rupununi R., Appun 2459 (K); Essequibo R., Ro. Schomburgk ser. I, 116 (K, NY, U). Suriname: Paramaribo Gardens, Stockdale s.n. (U, US); Paramaribo, van Hall 100 (U); Paramaribo, Agricultural Experiment Station, Samuels 8 (NY).

Uses: Same as those given for *Microtea debilis* (Ostendorf, 1962).

Note: Differences between the two *Microtea* species are most evident under magnification (see Fig. 1). The reticulate (honeycomb, waffle) pattern on the fruit of *M. debilis* exhibits protuberances which are very pale brown or cream, chartaceous, often irregularly bent or crisped, non-glochidiate at the apex, and to 0.2 mm long. By contrast, the (superficially) similar pattern on the fruit of *M. maypurensis* exhibits protuberances which are dark brown, hard (or at least stiff-looking), straight, glochidiate with 4 minute bristles at the apex, and to 0.5 mm long. The apex of the barb-like protuberance is provided with 4 retrorse, acicular bristles ca. 0.1 mm long, and an inner set of even shorter bristles. In addition, leaves from near the base or middle of the plant of *M. maypurensis* are often linear or narrowly elliptical, 0.3-1 cm wide, whereas leaves of the common *M. debilis* are usually wider (to 3 cm wide).

2. **PETIVERIA** L., Sp. Pl. 342. 1753.
Type: P. alliacea L.

Herbs or subshrubs, slightly woody at base. Leaves shortly petiolate; stipules present, minute. Inflorescences elongated, axillary or terminal spike-like racemes; flowers subsessile or pedicellate; bracteate and bracteolate. Flowers bisexual, more or less actinomorphic; tepals 4, free or partly united in a short tube, lobes spreading in flower, persistent, enlarging and erect in fruit; stamens 4-8, on a hypogynous disk, filaments unequal, shorter than tepals, anthers seemingly medifixed, linear; ovary 1-loculed, narrowly subellipsoid or oblongoid, flattened, pilose or tomentose, with 4-6 retrorse hook-like awns, ovule 1, style absent, stigma 1, papillose, penicillate and laterally decurrent along ventral (flattened) side of ovary. Fruits 2-lobed achenes with 4-6 apical hooked awns; seed 1, linear, testa membranous, adherent to pericarp, albumen scanty, mealy, cotyledons foliaceous.

Distribution: Species 2, occurring in tropical and subtropical America; 1 species (with 6 fruit-awns) endemic to Brazil, the other (with usually 4 fruit-awns) occurring in the United States, West Indies, Central and South America, including the Guianas.

1. **Petiveria alliacea** L., Sp. Pl. 342. 1753. Type: Herb. Clifford 141, *Petiveria* No. 1 (lectotype BM, not seen) (designated by Barrie in Jarvis *et al.* 1993: 74). – Fig. 2

Perennial erect herb to 1-2 m tall, with odor of garlic or onion when crushed. Stem angled or ribbed, puberulent in lines between ribs. Leaves elliptical, obovate or ovate, to 16(-20) x 7 cm, acute to acuminate (sometimes obtuse to subretuse, de Granville 2532) at apex, with minute bristle-tipped enations on margin, densely but very minutely pellucid-punctate, lower surface glabrous or sparingly puberulent on veins; petiole to 2 cm long, glabrate; stipules linear. Inflorescence a laxly flowered spike-like raceme to 45 cm long; rachis puberulent; flowers sessile or subsessile; bracts lanceolate to deltate, 1-3 mm long, puberulent, green; bracteoles persistent, 1 mm long. Tepals linear or oblong, 2.6-5 x 0.8-1 mm, acute or acuminate at apex, often pubescent at base and on lower part of veins, prominently 3- to 5-veined, spreading and white or pink in flower, erect and greenish in fruit; stamens free, unequal, to 3 mm long, filaments subulate, pink, anthers seemingly medifixed, linear, ca. 1.4 mm, deeply sagittately cleft at base and apex; ovary densely pubescent. Fruit linear or narrowly oblongoid, 8-10 mm long, cuneate, striate, puberulent, appressed to inflorescence-axis, with 4 (-6, de Granville 2532) apical hooked, 2.5-4 mm long awns.

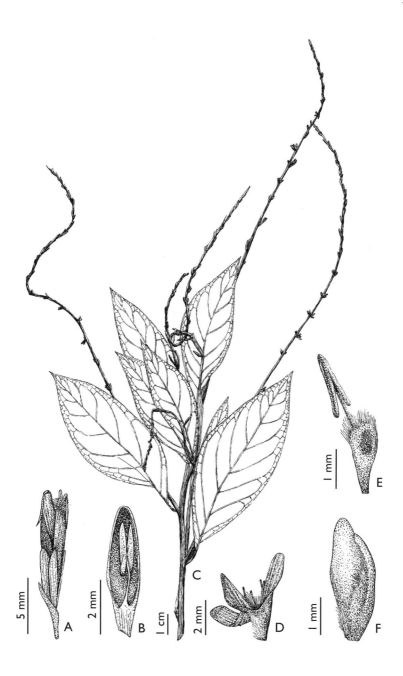

Fig. 2. *Petiveria alliacea* L.: A, awned fruit with bracts; B, stamen; C, habit; D, tepals; E, stamen; F, fruit. Drawing by Paul Pardoen.

10

Distribution: Southern United States, West Indies, Central and South America; in the Guianas a weed in disturbed areas such as roadsides and flower beds; 32 collections studied, all from the Guianas (GU: 5; SU: 4; FG: 23).

Selected specimens: Guyana: Emprensa, Rupununi, Gorinsky 21 (K); Georgetown, Promenade Gardens, Hitchcock 16608 (US). Suriname: Nieuw-Nickerie, Hekking 1124 (K, U); Tibiti savanne, Lanjouw & Lindeman 1814 (U). French Guiana: Saül, Mori et al.18246 (NY); Ile de Cayenne, Burgot 4 (CAY).

Uses: In Suriname, the plant is placed in chicken coops to help rid the fowl of lice, or an infusion of the plant is smeared on the birds for the same purpose (Ostendorf, 1962). In Guyana, as noted by Lachman-White et al. (1987), the entire plant is boiled, normally with minnie-root (*Ruellia tuberosa*), coconut (*Cocos nucifera)* root and pigeon-pea (*Cajanus cajan*) leaves, and the liquid used as a purgative by women suffering from "bladder troubles" (leucorrhoea). The plant is boiled with minnie-root, st. john's-bush (*Justicia secunda*) and inflammation-bush (*Vernonia cinerea*) and small quantities of the decoction are drunk for the relief of menstrual pains; larger doses are said to procure abortion. A decoction of the plant is also used as a tonic "for female rejuvenation" and as a diuretic. In French Guiana, the roots are used as an antispasmodic and febrifuge, and the leaves are decocted as a sudorific (Grenand et al., 1987). In French Guiana, the plant is locally used in sorcery to expel evil spirits (note on Burgot 4), and to bring good luck (Gely 53); a decoction is used to wash the walls of a house to protect it from evil spirits (Oldeman B-3892). Also in French Guiana the plant is used as an ingredient in a bath to treat fever (Grenand 2875).

Vernacular names: Guyana: gully-root; mucuraka. French Guiana: dongo dongu (Taki-taki); douvadouva, douvan-douvan (Creole); kananumna (Palikur); maipouri (Creole); ndongu-ndongu (Boni).

3. **PHYTOLACCA** L., Sp. Pl. 441. 1753.
Lectotype: P. americana L.

Herbs, shrubs or rarely trees. Leaves sessile or petiolate; stipules absent. Inflorescences mostly terminal racemes or spikes, occasionally thyrsiform or paniculiform below; bracts 1-3, at base of pedicel; bracteoles randomly inserted along pedicel. Flowers bisexual, or unisexual and dioecious; sessile or pedicellate; tepals 5, free, equal or subequal; stamens (5-)8-22(-33), in 1 or 2 whorls, on a hypogynous disk;

ovary 5-16-carpellate, carpels free or united, ovule 1 per carpel, basal, campylotropous, styles free, spreading or appressed-recurved. Fruits aggregations of 1-seeded drupelets ("berries") or syncarpous 5-16-loculed berries; seed 1 per cell, reniform or subreniform, black, shining, endosperm mealy, embryo curved or annular.

Distribution: Approximately 26 species, nearly cosmopolitan and with most species in the New World tropics and subtropics; 2 species in the Guianas.

KEY TO THE SPECIES

1 Inflorescence a simple raceme (rarely 1 or 2 branches at base); pedicels glabrous . . . . *1. P. rivinoides*
Inflorescence weakly subthyrsiform, with at least the lower flowers in groups of 2 or 3; pedicels weakly to strongly puberulent . . . . *2. P. thyrsiflora*

1. **Phytolacca rivinoides** Kunth & C.D. Bouché in Link, Kunth & C.D. Bouché, Ind. Sem. Hort. Bot. Berol. 1848: 15. 1849. Type: Venezuela, Caracas, Moritz s.n. (B, not seen, presumed destroyed).   – Fig. 3

Woody herb or subshrub to 3(-5) m tall. Stem ribbed, angled, glabrous. Leaves ovate, elliptical, lanceolate or oblong, to 21 x 8 cm, apex acute or acuminate, base acute or obtuse, glabrous, with pink-tinged veins, calcium oxalate raphides in epidermis visible with lens; petiole glabrous, 1-7 cm long. Inflorescence a many-flowered terminal or (pseudolateral) axillary raceme to 65(-70) cm long, to ca. 4 cm wide, simple (or rarely with 1-2 long branches from near base), much longer than leaves, rachis turning bright purple, glabrous; flowers widely spaced; bracts subtending pedicel linear, lance-linear or lanceolate, 1.5-5 mm long, hyaline; bracteoles 1-2, inserted randomly along pedicel but sometimes paired or subopposite, linear, 0.4-1.5 mm long; pedicels glabrous, pink, slender, 5-17 mm long. Tepals ovate, 1.2-2.8 x 1.5 mm, obtuse, white or pink, membranous, glabrous, becoming reflexed, deciduous in fruit; stamens (7-)9-14(-22), filaments 1.5-2 mm, anthers elliptical, 0.8 mm, cordate at base and apex; ovary of 9-16(-18) connate carpels, styles 10-16, slender, appressed-recurved, white. Berry becoming dark purple or black, with red juice, depressed globose, 6-8 mm diam., when dry 2-8 mm wide, ribbed at carpel-margins; seeds 9-18 (1 seed per carpel), ovoid, 1-2.6 x 1.3-2 mm, biconvex, black, shining.

12

Fig. 3. *Phytolacca rivinoides* Kunth & C.D. Bouché: A, fruit; B, seed; C, habit; D, ovary; E, stamen. Drawing by Paul Pardoen.

Distribution: New World tropics (West Indies, Mexico south to Bolivia, Argentina and southern Brazil); in the Guianas found along trails, roadsides, and altered ground such as airstrips and disturbed forests; 121 collections studied, all from the Guianas (GU: 41; SU: 34; FG: 46).

Selected specimens: Guyana: Northwest Distr., Mabaruma Compound, Archer 2295 (US); Mazaruni Station, Sandwith 1566 (K); Marudi Mts., Mazoa Hill, near Norman Mines camp, Stoffers *et al.* 193 (MO, U). Suriname: Marowijne R., Hugh-Jones 25 (K); Granmanzwamp, Saramacca, Reijenga 975 (U). French Guiana: Saül, trail to Limonade Cr., Mori *et al.* 19065 (NY).

Uses: Often cultivated in Suriname for the leaves, which are boiled and eaten as a vegetable (Ostendorf, 1962 and herbarium label of Hugh-Jones 25). As noted by Lachman-White *et al.* (1987), in Guyana "The leaves and young stems are eaten as a remedy for diabetes and a decoction of roots is drunk as a treatment for syphilis." Also, in Guyana the leaves are eaten raw in salad or cooked (label of Kvist 5). Further noted as consumed as a green-leafed vegetable in Guyana (label of Archer 2295) and French Guiana (label of Mori *et al.* 21186).

Vernacular names: Guyana: buck bhajee, deer kalaloo. Suriname: blaka-wiwiri, gogomago, makoko. French Guiana: bichouillac; macoco (Taki-taki).

Notes: The number of stamens per flower is variable; our observations include 15-22 (Gleason 194), 12-20 (Gleason 55) and 9-10 (Gleason 826). Occasionally a second flower, sessile or with a glabrous pedicel, will be formed upon the main pedicel of a *P. rivinoides* flower, but the phenomenon never extends to the condition of numerous multi-pedicels that pertains in *P. thyrsiflora*, which has pubescent pedicels; specimens of such *P. rivinoides* are Jansen-Jacobs *et al.* 1546 and de la Cruz 4481 from Guyana, and Jacquemin 1467 from French Guiana.
Normally *P. rivinoides* has glabrous pedicels, but occasionally they are densely beset with the sessile gland-like dots or excrescences which are typical of pedicels of *P. thyrsiflora*; such specimens (e.g., Hahn 5676 from Guyana) are, however, without the 2- to 3-flowered and - pedicellate axes of *P. thyrsiflora*. In *P. thyrsiflora* some of the gland-like dots or excrescences are sometimes produced into sparsely puberulous areas of trichomes to 0.5 mm.

Dried specimens may superficially resemble *Rivina humilis*; both taxa have glabrous pedicels, but the pedicel-bracteoles of *R. humilis* directly subtend the flower, whereas the pedicel-bracteoles of *Phytolacca rivinoides* are inserted randomly along the pedicel, sometimes near the middle and paired. However, this species has leaves with a rounded or truncate, not acute to obtuse, base; in the field, with its green inflorescence and orange to red-purple, globose berries, confusion is not possible.

2. **Phytolacca thyrsiflora** Fenzl ex J.A. Schmidt in Mart., Fl. Bras. 14(2): 343. 1872. Type: not designated.

Shrub to 3 m tall. Branches glabrous. Leaves ovate, lanceolate, oblanceolate, or lance-elliptical, ca. 7.5-20 x 3-9 cm, acuminate, often undulate and decurrent to base, glabrous; petiole sulcate, 1.5-4 cm long. Inflorescence a terminal or pseudolateral, many-flowered, weakly subthyrsiform raceme (rarely a panicle with 2 branches from base, Sastre 282) 9-30 cm long, longer than leaves, axes pink; bracts linear, 2-6 mm long; pedicels weakly to strongly puberulent, to 15 mm long, lower ones with 1-2 lateral flowers; bracteoles linear, 1-2.5 mm long. Tepals greenish-white to pink, concave, elliptical, 2-4 x 1.5-2 mm, sessile-glandular dotted on middle of back, reflexed in fruit; stamens 9-12, outer ones often staminodial, in 2 series on a hypogynous disk; ovary of 7-8 connate carpels, styles slender, recurved. Fruit depressed, becoming dark purple to black, ca. 5-7 mm wide; seed reniform, ca. 2 x 2 mm, smooth, shining, black.

Distribution: Venezuela, Guyana, French Guiana and Brazil; 43 collections studied, 17 from savanna woodlands in the Guianas (GU: 9; FG: 8).

Selected specimens: Guyana: Demerara-Mahaica Region, along Soesdyke-Linden Highway from Timehri Airport to Kuru-Kuru Cr., Hahn 3903 (U, US); road from Linden to Ituni, ter Steege *et al.* 224 (U); Mazaruni-Potaro Region, Pakaraima Mts., headwaters of Mazaruni R., banks of Mazaruni W of Imbaimadai, Pipoly 7924 (U). French Guiana: 1 km des Hates, village Indien, embouchure du fleuve Mana, Sastre 282 (CAY); Vicinity of Cayenne, Matabou Hill, Brants Road, Broadway 737 (NY,US); Mt. Mahury, 12 km E de Cayenne, Oldeman 1253 (US).

Vernacular name: Guyana: deer kalaloo.

Note: Lemée (1955), cites a French Guiana (Maroni, Camp Godebert) specimen of *Phytolacca americana* L. (*P. decandra* L.), but we have seen no material of it from the Guianas.

4. **RIVINA** L., Sp. Pl. 121. 1753.
   Type: R. humilis L.

Annual herbs or subshrubs, woody at base; stems erect or sprawling. Leaves petiolate; stipules absent. Inflorescences many-flowered, axillary or terminal racemes. Flowers bisexual, pedicellate; tepals 4, subequal, elliptical or obovate-oblong, obtuse or acute at apex, persistent and erect or spreading in fruit; stamens (3-)4(-9), alternating with tepals, filaments free, anthers dorsifixed; ovary 1-carpellate, ovoid or subglobose, ovule 1, style 1, subterminal, curved, short but distinct, stigma capitate or slightly lobed. Fruit a globose drupe; seed 1, lenticular, often pubescent, embryo curved, testa crustaceous.

Distribution: One very variable species, indigenous to tropical and subtropical America, including the Guianas.

1. **Rivina humilis** L., Sp. Pl. 121. 1753. Lectotype: Herb. Clifford 35, *Rivina* No. 1 (BM, not seen). – Fig. 4

Annual herb or subshrub to 1.2 m tall, often shrubby and with spreading, vinelike branches from base. Stem ribbed, glabrous to pubescent. Leaves lanceolate, ovate-lanceolate, ovate or oblong, sometimes elliptical, to 15 x 9 cm, entire or with low teeth, acute or acuminate at apex, rounded to truncate at base, glabrous or pubescent, often undulate; petiole to 11 cm long, canaliculate, pubescent with stiff hairs continuing on midvein of lower leaf surface. Inflorescence an axillary or terminal raceme to 20 cm long, axes glabrous or pubescent; bracts very narrowly lance-linear, 0.6-1 mm long, membranous, ciliate especially at mid-margin; bracteoles scalelike, immediately subtending perianth, <0.3 mm long; pedicels glabrous, 2.4-4 mm long in flower, elongating to ca. 7 mm in fruit. Tepals 4, obovate to linear-oblong or oblong-elliptic, 2.0-3.5 x 1 mm, obtuse, glabrous or ciliate, white, greenish or pinkish, partially enclosing fruit; stamens 4, ca. 1.3-1.5 mm long, anthers 0.5-1 mm; ovary ca. 1 mm, style 0.5 mm. Fruit ovoid, 2.5-4.5(-5) mm wide, smooth, red (sometimes orange or purple), black when dry; seed 2.5-3.5 mm, bluish-black, glabrous or hispid.

Distribution: New World from southern United States and West Indies to Argentina, now a pantropical weed, in the Guianas a weed in disturbed areas; 11 collections studied (GU: 2; SU: 7; FG: 2).

Fig. 4. *Rivina humilis* L.: A, flower with bracteoles and tepals; B, seed; C, fruit; D, habit with inflorescence; E, flower with perianth removed. Drawing by Paul Pardoen.

Selected specimens: Guyana: Crossing of Soldier's Cr. (Kunaballi Cr.) and road to St. Culbert's Mission, Robertson & Austin 291 (MO); Georgetown, Promenade Gardens, Hitchcock 16587 (NY, US). Suriname: Peperpot Plantation opposite Paramaribo, Hekking 1164 (K, NY, U); Wia-Wia Reservation, Sterringa LBB 12424 (U); Bigi Santi, E of Mot Cr., Smith-Naarendorp LBB 12915 (U). French Guiana: Marche de Cayenne, Alexandre 338 (CAY); Cayenne, Alexandre 164 (CAY).

5. **SEGUIERIA** Loefl., Iter Hispan. 191. 1758.
Type: S. americana L.

Trees, more or less scandent shrubs, or lianas. Young branches (except those immediately below inflorescence) often with paired thorns. Leaves petiolate; stipules seemingly present at base of older leaves, as paired thorns, or absent. Inflorescences axillary and terminal, few-flowered racemes to irregular, profusely flowering panicles; lower bracts sometimes leaf-like, membranaceous; bracteoles similar to but smaller than bracts, or absent. Flowers bisexual, pedicellate; tepals 5, reflexed in fruit, membranous, colored, glabrous, subequal; stamens ca. 15-65, irregularly inserted, anthers versatile, extrorse, linear, cordate at base and apex; ovary 1-locular, ovule 1, basal, campylotropous, ovary compressed and merging with winged style, stigma 1, papillose, decurrent on thicker margin of style. Fruit a samara with a dorsal wing, growing out from elongating style, basal part sometimes bearing lateral winglets; seed 1, globose or compressed-globose, testa coriaceous, embryo extremely curved.

Distribution: Species 6, principally in tropical South America, mostly in secondary forests outside the Amazonian region; 2 species in the Guianas.

Literature: Rohwer, J.G. 1982. A taxonomic revision of the genera Seguieria Loefl. and Gallesia Casar. (Phytolaccaceae). Mitt. Bot. Staatssamml. München 18: 231-288.

KEY TO THE SPECIES

1 Tepals drying straw-yellowish-brown; ovary and fruit (samara) drying yellowish-brown, bearing small, lateral winglets at base . . . *1. S. americana*
  Tepals drying dark brown; ovary and fruit (samara) becoming black on drying, without lateral winglets . . . . *2. S. macrophylla*

1. **Seguieria americana** L., Syst. Nat. ed. 10. 1074. 1759. Neotype: Guyana, Rupununi Distr., S of Lethem, Takutu R., Irwin 797 (US!) (designated by Rohwer 1982: 237).   – Fig. 5

Seguieria foliosa Benth., Trans. Linn. Soc. London 18: 236. 1839. Type: Guyana, Ro. Schomburgk ser. I, 661 (B, BM, F, G, K); photo Delessert Herbarium (US).

Liana or scandent shrub (or small tree) climbing to ca. 10 m, trunk to 14 cm diam. Branchlets slender, subterete, striate, glabrous. Leaves subcoriaceous, ovate or elliptical, to 15.5 x 8 cm, acuminate or obtuse at apex, rounded at base; petiole canaliculate, glabrous, 3-10 mm long; stipular thorns recurved, stout, to 20 mm long. Inflorescence an axillary or terminal, 15- to ca. 100-flowered, to 40 cm long panicle; axes angular, sparsely pubescent to lanate; pedicels 3-10 mm long, subterete, densely puberulent; bracts leaf-like at base of inflorescence and diminishing in size upwards; bracteoles triangular, concave, to 1.3 mm.

18

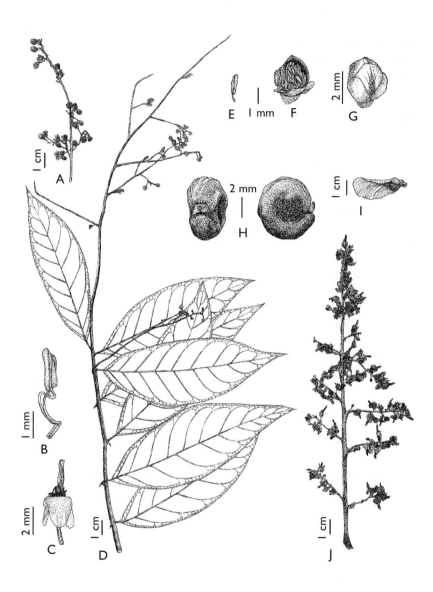

Fig. 5. *Seguieria americana* L.: A, inflorescence; B, stamen; C, pistil; D, habit; E, stamen; F, opened flower; G, tepals; H, seeds; I, winged fruit; J, infructescence. Drawing by Paul Pardoen.

Tepals ovate, to 7.5-8 mm, obtuse, white or pale yellow; stamens 30-65, filaments filiform, to 6.5 x 2 mm, anthers linear, to 2.4 mm; ovary ovoid, longitudinally ribbed, with distinct primordia of lateral winglets (fluted enations) at base, style erect, winged on one side, apex subrecurved. Samara to 5 cm long, basal part to 1.1 cm wide, with winglets up to 4 mm wide, terminal wing of variable elongated shape, to 2 cm wide; testa red-brown.

Distribution: The Guianas, Peru, and Brazil; 15 collections examined, all from the Guianas (GU: 5; SU: 7; FG: 3).

Selected specimens: Guyana: Kanuku Mts., Rupununi R., Puwib R., Jansen-Jacobs *et al.* 236 (B, NY,U); Rupununi Distr., S of Lethem, E bank of Takutu R., Irwin 797 (US); North Rupununi Savanna, Davis 893 (NY). Suriname: Distr. Nickerie, area of Kabalebo Dam project, Corentyne R., Heyde *et al.* 124 (U); Nat. Res. Raleigh Falls-Voltzberg, Foengoe Island, E of airstrip, Heyde 626 (U); Paramaribo, Wullschlägel 1683 (U). French Guiana: Rives du Yaroupi, environ 3 km en aval de Saut Tainoua, Oldeman 3151 (CAY); Trois Sauts, Abatis Alasuka, Grenand 1117 (CAY); Zidockville, Haut Oyapock, Prévost *et al.* 892 (CAY).

Uses: Based on the Prévost & Grenand and Grenand specimens from French Guiana cited above, Grenand *et al.* (1987) note that the Wayapi Indians use the plant as a hunting poison, and that the root and stem have almost the same effect as curare made of *Strychnos guianensis* (LOGANIACEAE), which makes the plant worthy of future intensive study.

Vernacular names: Suriname: boslemmetje. French Guiana: citron sauvage.

Note: We follow Rohwer (1982) in considering *Seguieria foliosa* Benth. conspecific with *S. americana* and different from *S. aculeata* Jacq. Nowicke (1968) treats the former as a valid species from Brazil and Guyana which shares many overlapping technical characteristics with *S. americana* and the latter as synonymous with *S. aculeata*.

2. **Seguieria macrophylla** Benth., Trans. Linn. Soc. London 18: 235. 1839. Type: Guyana, Ro. Schomburgk ser. I, 348 (K, not seen).

Tall liana or climbing shrub. Internodes of older branches hollow. Leaves subcoriaceous, elliptical or ovate-elliptical, to 18 x 8.5 cm, apex obtuse to acuminate (often also mucronulate), base rounded; petiole 3-14 mm long; stipular thorns recurved, stout, to 12 mm long.

Inflorescence an axillary or terminal, profusely (> 100) flowered panicle to 50 cm long, sparsely to densely puberulent; bracts acicular, to 5 mm; bracteoles acicular, to 1.5 mm; pedicels 3-9 mm long. Tepals 4.5(-6) x 3.5(-4.5) mm, greenish-yellow, becoming dark on drying; stamens ca. 15-40, filaments to 3.5 mm long, anthers 1.5-2.5 mm; ovary without lateral winglet primordia (enations), becoming black on drying, stigma normally completely lateral. Samara to 4 cm long, becoming black on drying, basal part to 8 mm wide, smooth or somewhat veined, terminal wing widened upwards to 1.6 cm and widest from middle to tip; testa black.

Distribution: Panama, Colombia, Venezuela, Trinidad, Guyana, French Guiana (?), Peru, and western and northern Brazil; 15 collections studied (GU: 3).

Selected specimens: Guyana: Northwest portion of Kanuku Mts., Mt. Iramaikpang, A.C. Smith 3650 (K, US); Espeiro, Corentyne R., Jenman 503 (K); Rupununi Distr., Kwitaro R., Brian's Landing, Jansen-Jacobs et al. 3835 (U, US).

Note: Lemée (1955) cites an incomplete, doubtful Leprieur specimen attributed to this species by Walter (1909).

6. **TRICHOSTIGMA** A. Rich. in Sagra, Hist. Fis. Pol. Nat. Cuba, Pt. 2, Hist. Nat. 10: 306. 1845.
Type: T. rivinoides A. Rich., nom. illeg. (Rivina octandra L., T. octandrum (L.) H. Walt.)

Woody shrubs or vines (lianas), rarely tree-like. Leaves alternate or subopposite, petiolate; stipules minute, deciduous. Inflorescences many-flowered terminal or pseudo-axillary, condensed or lax racemes; bracts linear, often inserted at or near middle of pedicel, deciduous; bracteoles 2, immediately subtending flower, persistent, minute. Flowers bisexual; tepals 4, subequal, persistent and reflexed in fruit; stamens 8-12(-25), filaments free, anthers basifixed; ovary 1-carpellate, 1-locular, cylindrical to globose, ovule 1, style short or absent, stigma often penicillate. Fruit a 1-seeded drupe, globose or lenticular, reddish-purple to black, pericarp adherent to seed; testa crustaceous.

Distribution: Species 3(-4), occurring in the New World tropics and subtropics (Florida, West Indies, Central and South America); 1 species in the Guianas.

1. **Trichostigma octandrum** (L.) H. Walt. in Engl., Pflanzenr. IV. 83 (Heft 39): 109, f. 31. 1909. – *Rivina octandra* L., Cent. Pl. 2: 9. 1756. Type: Herb. Linnaeus No. 163.3 (LINN, not seen).   – Fig. 6

Woody shrub or liana to 10 m; bark greyish-purple. Branches long, slender, minutely puberulent. Leaves oblong, elliptical or lanceolate (rarely ovate), 5-15 x 2-6 cm, acute or acuminate at apex, cuneate to acute or rounded at base; petiole canaliculate, sparsely pubescent, 1-3.7 cm long, glabrous, base of petiole persistent, peg-like on old stems. Inflorescence a laxly to rather densely many-flowered raceme to 13 cm long, often pseudo-axillary on very short, leafless shoots; pedicels 3-10 mm long; bracts lance-linear, 1-2 mm long, deciduous; bracteoles 2, scale-like, triangular, ca. 0.5 mm long, immediately below perianth. Tepals ovate or obovate, 3-5 x 2.5-3 mm, concave, obtuse, greenish-white or white in flower, turning red, purple or purplish-pink in fruit; stamens 8-12, free, anthers basally sagittate and deeply incised at apex; ovary subglobose, glabrous, style very short, stigma 1. Fruit subglobose, 4.5-5.5 mm wide, red, purple, black or brown; seed black, shining, 4-6 mm.

Distribution: New World tropics, from southern Florida and West Indies, to southern Brazil; in Guyana and French Guiana (see note); 55 collections studied (GU: 3).

Selected specimens: Guyana: Providence, Berbice, Jenman 5151 (K, NY, US); Demerara, Georgetown Botanic Garden, Fairchild s.n. (US); without locality, Jenman 2192 (NY).

Note: Lemée (1955), cites from French Guiana an Aublet specimen in the Denaiffe Herbarium, described by Aublet, Hist. Pl. Guiane 1: 90. 1775 under *Rivina octandra*.

22

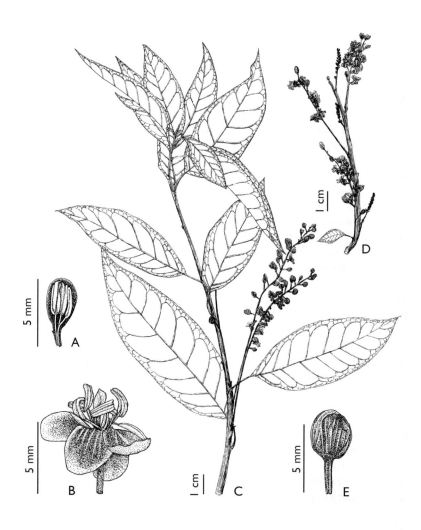

Fig. 6. *Trichostigma octandrum* (L.) H. Walt.: A, stamens; B, flower; C, habit; D, infructescence; E, tepals. Drawing by Paul Pardoen.

29. NYCTAGINACEAE

by

ROBERT A. DEFILIPPS AND SHIRLEY L. MAINA[3]

Annual or perennial herbs, shrubs or trees, sometimes scandent or prostrate; root system fibrous, fleshy or tuberous. Branches dichotomously or trichotomously forked, often swollen at nodes, sometimes armed with, and climbing by means of, thick, straight to recurved, axillary spines. Leaves alternate, opposite, subopposite or whorled, simple; sessile or petiolate; exstipulate; blade membranaceous to carneous, often marked and coarsened with conspicuous raphides, symmetric or oblique at base, margin usually entire, sometimes toothed or lobed, glabrous or pubescent; pinnately veined. Inflorescence of terminal or axillary, diffuse or congested, paniculate or corymbose cymes (sometimes flowers axillary and solitary, or in thyrses, racemes, spikes, umbels or capitules), pedunculate, usually bracteate or variously involucrate; involucre, when present, of free or connate segments, often calyx-like, enclosing 1 or many flowers, persistent or deciduous, often accrescent around fruit, green or brightly coloured. Flowers regular or nearly so, bisexual or unisexual, then sometimes sexes with different perianth, usually dioecious when flowers unisexual; perianth plicate or contorted in bud, green or coloured, campanulate, funnel-shaped, or tubular, sometimes urceolate or salver-shaped, connate completely to a truncate tube or for more than 2/3 with a 3-5-dentate or -lobed limb, upper part mostly caducous after anthesis, tube persistent in fruit, usually accrescent, indurate at base and forming an anthocarp; stamens 1-10(-30), hypogynous, included or exserted, filaments unequal, filiform, free or united at base in small, often swollen collar, presumably functioning as nectary, anthers dehiscent by longitudinal slits, 4-locular, 2-locular when mature, dorsifixed or basifixed; pistil 1, included in perianth, ovary superior, 1-carpellate, 1-locular, sessile or stipitate, ovule 1, anatropous or campylotropous, style short or elongate, sometimes absent, stigma simple, capitate, peltate, penicillate or fimbriate, terminal, rarely lateral or style stigmatic along one side. Fruit a 1-seeded anthocarp, formed by persistent, coriaceous, fleshy perianth-tube or its indurated base with enclosed achene or utricle; anthocarp costate (5- to 10-ribbed at base), sulcate or winged, glabrous or pubescent, often viscous when wet; seed 1, testa hyaline, endosperm scanty, perisperm abundant, usually mealy or fleshy, embryo curved or straight.

[3] National Museum of Natural History, Department of Systematic Biology - Botany, NHB 166, Smithsonian Institution, Washington, D.C. 20013-7012, U.S.A.

24

Distribution: About 400 species in 31 genera, mostly in tropical America, with a few temperate representatives; in the Guianas 14 species in 6 genera, including the cultivated *Bougainvillea* and *Mirabilis*.

LITERATURE

Bittrich, V. & U. Kühn. 1993. Nyctaginaceae. In K. Kubitzki, The Families and Genera of Vascular Plants 2: 473-486.

Bogle, A.L. 1974. The genera of Nyctaginaceae in the southeastern United States. J. Arnold Arbor. 55: 1-37.

DeFilipps, R.A. 1992. Ornamental Garden Plants of the Guianas: An Historical Perspective of Selected Garden Plants from Guyana, Surinam and French Guiana. Nyctaginaceae. pp. 161-163. Smithsonian Institution. Washington, D.C.

Dumas, M.A.D. 1991. Listado preliminar de Nyctaginaceas cubanas. Revista Jard. Bot. Nac. Univ. Habana 12: 23-25.

Fay, J.J. 1980. Flora de Veracruz. Nyctaginaceae. 13. 54 pp.

Heimerl, A. 1934. Nyctaginaceae. In H.G.A. Engler & K.A.E. Prantl, Die natürlichen Pflanzenfamilien, ed. 2. 16c: 86-134.

Jarvis, C.E., *et al*. 1993. A List of Linnaean Generic Names and their Types. Regnum Veg. 127. 100 pp.

Kellogg, E.A. 1988. Nyctaginaceae. In R.A. Howard, Flora of the Lesser Antilles 4: 173-186.

Lemée, A.M.V. 1955-1956. Flore de la Guyane Française. Nyctaginacées. 1: 570-574. 1955; 4: 32. 1956.

Maguire, B. *et al.*, 1948. Plant Explorations in Guiana in 1944, chiefly to the Tafelberg and the Kaieteur Plateau - III. Bull. Torrey Bot. Club 75: 286-323.

Nicolson, D.H. 1991. Flora of Dominica, Part 2: Dicotyledoneae. Nyctaginaceae. Smithsonian Contr. Bot. 77: 168-169.

Ostendorf, F.W. 1962. Nuttige Planten en Sierplanten in Suriname. Nyctaginaceae. Bull. Landbouwproefstat. Suriname 79: 26-27.

Pulle, A.A. 1906. An Enumeration of the Vascular Plants known from Surinam. Nyctaginaceae. pp. 169-170.

Reitz, P.R. 1970. Flora Ilustrada Catarinense. Nictagináceas. 52 pp.

Schmidt, J.A. 1872. Nyctagineae. In C.F.P. von Martius, Flora Brasiliensis 14(2): 345-376.

Standley, P.C. 1918. North American Flora. Allioniaceae. 21: 171-254.

Standley, P.C. 1931. The Nyctaginaceae of northwestern South America. Field Mus. Nat. Hist., Bot. Ser. 11: 73-114.

Standley, P.C. 1937a. Flora of Costa Rica 2. Nyctaginaceae. Publ. Field Mus. Nat. Hist., Bot. Ser. 18: 423-426.

Standley, P.C. 1937b. Nyctaginaceae. In J.F. Macbride, Flora of Peru 2. Field Mus. Nat. Hist., Bot. Ser. 13(2): 518-546.

Standley, P.C. & J.A. Steyermark. 1946. Flora of Guatemala. Nyctaginaceae. Fieldiana Bot. 24(4): 174-192.

Steyermark, J.A. 1987. Flora of the Venezuelan Guayana - III. Nyctaginaceae. Ann. Missouri Bot. Gard. 74: 614-635.

Woodson, R.E. & R.W. Schery. 1961. Nyctaginaceae. In Flora of Panama 4: 393-407 = Ann. Missouri Bot. Gard. 48: 51-65.

KEY TO THE GENERA

1 Herbs or subshrubs; stems unarmed; flowers bisexual; leaves opposite or subopposite 2
  Shrubs, trees or woody climbers; stems armed or unarmed; flowers bisexual or unisexual; leaves alternate, opposite or subopposite 3

2 Inflorescence a panicle; without involucre; perianth 0.7-1 cm long . . . .
  . . . . 1. *Boerhavia diffusa*
  Inflorescence a terminal cyme; calyx-like involucre coloured; perianth 3-5.5 cm long, corolla-like . . . . 4. *Mirabilis jalapa*

3 Leaves alternate; branches often with supra-axillary spines; flowers bisexual, perianth 1.5-3 cm long, in groups of 3 surrounded by an involucre of leafy, bright-coloured bracts . . . . 2. *Bougainvillea*
  Leaves opposite, subopposite or in part verticillate; branches unarmed or with axillary spines; flowers unisexual, perianth less than 1 cm long, bracts minute, not forming an involucre 4

4 Scrambling shrub or liana, often with axillary, slightly curved spines; perianth and anthocarp with stipitate glands . . . . 6. *Pisonia macranthocarpa*
  Shrubs or trees, unarmed; perianth and anthocarp without glands 5

5 Perianth of male flowers obconic-campanulate; stamens exserted; female flower buds rounded or obtuse at apex . . . . 3. *Guapira*
  Perianth of male flowers urceolate, globose or elongate; stamens included; female flower buds long-acuminate at apex . . . . 5. *Neea*

N o t e : Since specimens of *3. Guapira* and *5. Neea* are sometimes difficult to separate, a combined key covering Guianan species of both genera is provided below, as well as a key to the species within each genus which is located in the respective generic treatments.[4]

_____

[4] We are deeply grateful for the assistance of J.C. Lindeman and L.Y.Th. Westra in developing the combined key and to A.L. Stoffers for generously supplying much other data.

COMBINED KEY TO VEGETATIVE CHARACTERISTICS OF SPECIES OF
3. Guapira and *5. Neea*

1 Leaves glossy above, usually less than 12 x 5 cm, lateral veins 15-25 per side, 3-4 mm apart . . . . *5-4. N. ovalifolia*
 Leaves usually not glossy above, lateral veins mostly less than 15 per side, always more than 4 mm apart 2

2 Branches, petioles and leaf underside with ferruginous hairs (loupe) . . . 3
 Branches, petioles and leaves glabrous or nearly so, except the youngest parts 7

3 Hairs on vegetative parts spreading 4
 Hairs on vegetative parts more or less appressed, minute 5

4 Leaves glabrous above . . . . *5-3. N. mollis*
 Leaves villous above, especially on midrib . . . .
 . . . . *G. marcano-bertii* (Venezuela; see note to 3-3. *G. kanukuensis*)

5 Leaves drying brownish 6
 Leaves drying blackish-brown . . . .
 . . . *3-3. G. kanukuensis* (and see note to *3-3*: Venezuelan *G. amacurensis*)

6 Leaves ovate, apex cuspidate-acuminate, often rufous beneath . . . .
 . . . . *3-1. G. cuspidata*
 Leaves elliptical, apex acuminate, not cuspidate, both sides same colour, usually 3 times as long as wide . . . . *3-4. G. salicifolia*

7 Leaves elliptical, lateral veins obscure, 5-6 per side 8
 Leaves often widest above middle, lateral veins prominent beneath, 8 or more per side 9

8 Leaves asymmetrical, midrib slightly curved . . . . *5-1. N. constricta*
 Leaves more or less symmetrical . . *3-2. G. eggersiana* (incl. *G. guianensis*)

9 Lateral veins 8-10 per side, anastomosing 10-15 mm from margin, impressed above; tree cauliflorous . . . .
 . . . . *N. davidsei* (Venezuela; see note to 5-1. *N. constricta*)
 Lateral veins 10-18 per side, anastomosing less than 5 mm from margin, prominulous above 10

10 Lateral veins arcuate, smaller ones irregular, rather obscure; tree mostly cauliflorous . . . . *5-2. N. floribunda*
 Lateral veins nearly straight to the loop with smaller ones in between, smaller veins distinct and prominulous beneath; tree not cauliflorous
 . . . . *5-5. N. spruceana*

1. **BOERHAVIA** L., Sp. Pl. 3. 1753.
Lectotype: B. erecta L.

Annual or perennial herbs, sometimes suffrutescent or subscandent; glabrous or pubescent. Leaves opposite or subopposite, those of a pair often unequal, large leaves often alternating; petiolate; blade entire, sinuate or undulate, raphides prominent, linear, base sometimes inequilateral. Inflorescences varying from a terminal or axillary panicle to paniculate capitula, or flowers arranged in racemes, unequally pedicellate umbels or cymes, flowers rarely solitary; branches often subtended by a small hyaline bract; pedicellate or subsessile; bracteoles 1-3, minute, hyaline. Flowers bisexual, small; perianth corolla-like, campanulate, subrotate or funnelform, lower part constricted above ovary, persistent, upper part (limb) shallowly 5-lobed, flaring, deciduous; stamens 1-3(-5), exserted or included, filaments free or connate below, pollen forate, 2(-3)-nucleate; ovary shortly stipitate, ovule anacampylotropous, style filiform, stigma 1, capitate, subexserted to exserted. Anthocarps shortly stipitate, clavate, fusiform or subellipsoid, 3-5(-10)-angled, -ribbed or -winged, green, with stipitate, viscous glands on ribs (eglandular in *B. erecta*, not in the Guianas), and streaks of white, linear superficial raphides in furrows between ribs; seed with thin testa adherent to pericarp, embryo with unequal cotyledons curved inward, enclosing scant, mealy endosperm, radicle elongate, descending.

Distribution: About 20 species of pantropical and subtropical weeds, 1 of these in the Guianas.

Literature: Fosberg, F.R. 1978. Studies in the genus Boerhavia L. (Nyctaginaceae), 1-5. Smithsonian Contr. Bot. 39: 1-20.

1. **Boerhavia diffusa** L., Sp. Pl. 3. 1753, nom. cons. prop. Type: Virgin Islands, St. Croix, Teague Bay, Fosberg 56776 (BM, not seen), typ. cons. prop.   – Fig. 7

Boerhavia paniculata Rich., Actes Soc. Hist. Nat. Paris 1: 105. 1792. Type: French Guiana, Cayenne, Leblond s.n. (P, holotype, not seen).
Boerhavia surinamensis Miq., Linnaea 18: 244. 1844. Type: Suriname, Nickerie, Focke 500 (not seen).
Full synonymy (including many names not ascribed to Guianan plants) is provided in Woodson & Schery (1961).

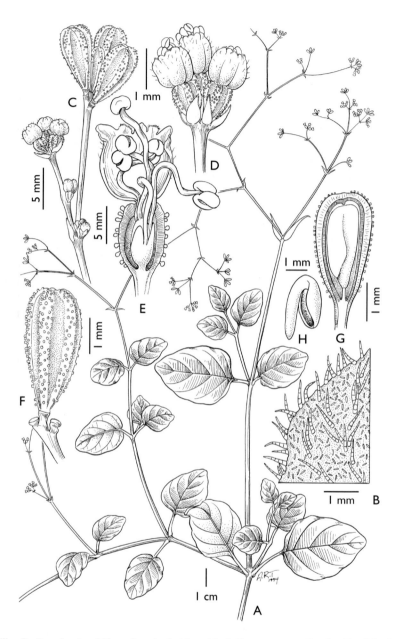

Fig. 7. *Boerhavia diffusa* L.: A, habit with inflorescences; B, detail of leaf trichomes and raphides; C, detail of inflorescence and infructescence; D, flowers; E, section through flower; F, fruit; G, section through fruit; H, seed (A-E, G-H, Westra 10533, Dunlap 377; F, Maas *et al.* 7599). Drawing by A.R. Tangerini.

Perennial herb; taproot woody, fusiform, to 1 cm wide. Stem prostrate to decumbent or ascending, to 2 m long, slender, sparsely branched, widely spreading, glabrous or puberulent. Petiole to ca. 4.3 cm long; blade orbicular, rhombic-orbicular or rhombic-ovate on lower part of stem, frequently intergrading to ovate or lanceolate on upper part, 1.1-6 x 0.8-5 cm, apex obtuse to acute, base cuneate, truncate or obtuse, margin sometimes undulate, sub-scalloped or sinuate, and ciliate with multicellular, to 1 mm long hairs, upper surface yellowish-green, glabrous, lower surface paler, glabrous or with puberulent veins. Inflorescence a terminal or axillary, pedunculate, strict or diffusely branched, to 45 cm long panicle, bearing few- to many-flowered subumbellate clusters of subsessile or pedicellate flowers; branches glabrous or glandular-puberulent, bracteolate at base; bracts lanceolate, hyaline; pedicel 0-0.5 mm long; bracteole 0.9-1.7 mm long. Perianth red, purplish-red, maroon or purple, tube closely adherent to ovary, 0.7-1.0 x 0.5 mm, 4-5-ribbed, minutely glandular on ribs, limb broadly campanulate, 5-lobed, sometimes puberulent and ciliate, limb including lobes ca. 1.3 x 0.5 mm, lobes ca. 0.5 mm; stamens 1-3, included to slightly exserted, filaments recurved, united at base; ovary narrowly ellipsoid, ca. 0.5 mm wide, style recurved, stigma slightly exserted, peltate. Anthocarp sessile or subsessile, clavate, narrowly oblanceoloid or subellipsoid, to 4 x 1.5 mm, obtuse, green, 4-5-ribbed, ribs stipitate-glandular, viscous, groove between ribs smooth, with numerous linear, white raphides appearing as minute streaks; seed narrowly ovoid, ca. 2 mm, castaneous.

Distribution: Pantropical and subtropical, sometimes extending to warm temperate zones such as southern California (U.S.A.); in the Guianas a weed, usually found in open, sandy or stony soil on roadsides; 72 collections studied, all from the Guianas (GU: 19; SU: 36; FG: 17).

Selected specimens: Guyana: Rupununi R., Monkey Pond landing, SW of Mt. Makarapan, Maas *et al.* 7599 (B, CAY, NY, U, US). Suriname: Experimental Farm Kabo, Saramacca Distr., Everaarts 567 (CAY, U). French Guiana: Bac de Mana, Bassin de Basse-Mana, Cremers *et al.* 10494 (CAY, US).

Use: French Guiana: Root is emetic and purgative.

Vernacular name: French Guiana: ipecacuanha de Cayenne.

Note: Woodson & Schery (1961), in treating this species for the Flora of Panama, observed no tangible taxonomic differences between the two major entities contesting within it for recognition: (1) *B. coccinea* Mill.

(including *B. caribaea* Jacq.) of the New World (usually described as having inflorescence-branches puberulent; capitula variously designated as 4-20-flowered (by Nicolson, 1991) or 6-12-flowered (by Kellogg, 1988)); and (2) *B. diffusa* (including *B. paniculata*) of the Old World (inflorescence-branches glabrous, ebracteate; capitula variously designated as 1-5-flowered (by Nicolson, 1991) or 2-4(-7)-flowered (by Kellogg, 1988)), endemic to Sri Lanka and possibly southern India according to Fosberg (1978). The present authors concur with the disposition by Woodson & Schery (1961) of such variation within a single species, *B. diffusa*. Such opinion is also in accord with that of Bogle (1974), who referred to the "morphologically variable pantropic weed *Boerhavia diffusa* L. (incl. *B. caribaea* Jacq., *B. coccinea* Mill., *B. decumbens* Vahl, *B. hirsuta* Willd., *B. paniculata* Rich., *B. viscosa* Lag. & Rodr.)".

2. **BOUGAINVILLEA** Comm. ex Juss., Gen. Pl. 91. 1789, nom. et orth. cons.
Type: B. spectabilis Willd.

Shrubs or trees; often with long, clambering branches, sometimes armed with simple or apically forked, supra-axillary spines, glabrous or pubescent. Leaves alternate; petiolate; blade broad, entire. Inflorescences axillary or terminal cymes with pseudanthia consisting of an involucre of 3 persistent, leafy, coloured bracts, each (in the Guianas) bearing a bisexual flower with its pedicel adnate to the midvein. Perianth tubular, tube subterete or 5-angled, limb 5(rarely 4)-lobed, lobes induplicate-valvate; stamens (4-)5-10, included, filaments somewhat unequal, capillary, connate at base in a short cup; ovary stipitate, fusiform, slightly laterally compressed, style subterminal, short, filiform or subclavate, straight or slightly curved, included, partially or completely papillose. Anthocarps fusiform, coriaceous, 5-costate; seed with thin testa adherent to pericarp, embryo uncinate, cotyledons incumbent, enclosing mealy endosperm, radicle descending.

Distribution: Approximately 18 species, occurring in Central America and tropical South America; 2 species cultivated in the Guianas.

Note: Commerson (1727-1773) discovered and collected a specimen of Bougainvillea in Brazil, while accompanying the explorer Louis Antoine de Bougainville (1729-1811) on a portion of his circumnavigation of the world in 1766-1769. Commerson suggested that the new genus be named after Bougainville, and when specimens from the trip had been sent to A.-L. de Jussieu in Paris, the latter botanist described the genus in 1789.

Literature: Gilbert, R. 1992. Brilliant bracts. Garden (London 1975+) 117: 106-109. MacDaniels, L.H. 1981. A study of cultivars in Bougainvillea. Baileya 21: 77-100.

KEY TO THE SPECIES

1 Leaves and stems glabrate; perianth tube distinctly 5-angled below, glabrate or puberulent with antrorsely curved, to 0.5 mm long hairs . . . *1. B. glabra*
  Leaves and stems villous; perianth tube veined but not distinctly angular below, densely pubescent with spreading, straight, to 1 mm long hairs . . .
  . . . . *2. B. spectabilis*

1. **Bougainvillea glabra** Choisy in DC., Prodr. 13(2): 437. 1849. Lectotype: Brazil, Rio de Janeiro, Gaudichaud 423 (G-DC; IDC 800. 2215: III. 2, photo, not seen) (designated by Kellogg, 1988). – Fig. 8

Bougainvillea spectabilis Willd. var. *glabra* (Choisy) Hook., Bot. Mag. 80: ad t. 4811. 1854.

Woody vine, armed, high-climbing, or tree to 20 m; branches yellowish or reddish-brown, puberulent when young, but soon glabrate; spines 2-15 mm long, stout, straight or recurved. Petiole 0.6-2.3 cm long, sparsely puberulent; blade broadly ovate to elliptic-ovate or ovate-lanceolate, 3-10 x 1.8-5.3 cm, abruptly or gradually acute to long-acuminate at apex, rounded to acute at base, puberulent when young, soon glabrate. Pseudanthia 3-flowered, peduncle 1-2.5 cm long; bracts pink, purple, purplish-red or magenta, elliptical to broadly ovate, 2.4-5.2 x 1.7-3.9 cm, obtuse to abruptly acute or acuminate at apex, cordate at base, sparsely puberulent or glabrous. Perianth 1.5-2.5 cm long, tube greenish or yellowish-white, glabrate or puberulent with antrorsely curved, to 0.5 mm long hairs, distinctly obtusely 5-angled below; stamens 8. Anthocarp turbinate, 7-13 x 4-5 mm, acutely 5-angled, glabrate or sparsely puberulent.

Distribution: Eastern and southern Brazil; widely cultivated in tropical and subtropical areas as an ornamental, including the Guianas; 1 collection studied from Guyana: Demerara-Mahaica Region, campus of the University of Guyana, cultivated, Hahn 4850 (US).

Vernacular name: Suriname: bougainville.

32

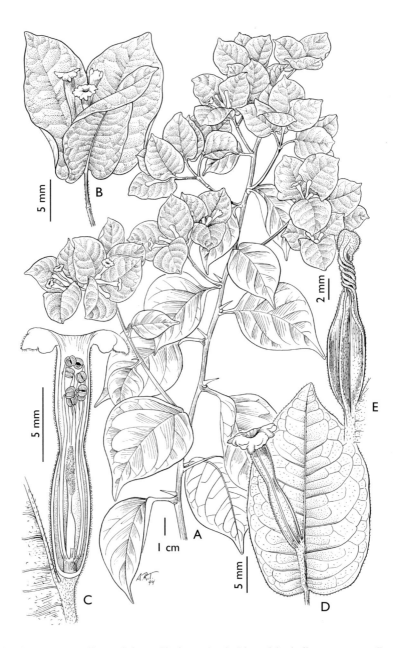

Fig. 8. *Bougainvillea glabra* Choisy: A, habit with inflorescences; B, pseudanthium with bracts and flowers; C, section through flower; D, young fruit and subtending bract; E, mature ribbed fruit (A-E, Baker 1908). Drawing by A.R. Tangerini.

2. **Bougainvillea spectabilis** Willd., Sp. Pl. 2: 348. 1789. Type: Brazil, Humboldt s.n. (herb. Willd. #7332; IDC 7440.504:II.5, photo, not seen).

Woody vine, armed, high-climbing, or tree to 12 m; branches flexuous, greyish or reddish-brown, usually copiously fulvous-villous, sometimes glabrate; spines to 4 cm long, stout, straight or curved. Leaves petiolate; blade broadly ovate to suborbicular or rounded-ovate, 5-10 x 2.5-6.5 cm, abruptly acute or acuminate at apex, rounded to acutish at base and often short-decurrent, sparsely short-villous above, often glabrate in age, usually densely villous beneath. Pseudanthia 3-flowered, peduncle 6-17 mm long, slender; bracts pink or purplish-red, ovate or broadly ovate, 2-5 x 1.7-3.1 cm, abruptly acute or acuminate or sometimes obtuse at apex, subcordate at base, sparsely puberulent or short-villous. Perianth 1.5-3 cm long, tube green, densely short-pubescent with spreading, straight, to 1 mm long hairs, lower part veined but not prominently angled below, limb 6-7 mm wide, lobes ovate-triangular, subobtuse, yellowish above; stamens 7-10, often 8. Anthocarp oblong-ellipsoid, 11-14 x 5 mm, 5-ribbed, greyish-green, densely short-villous.

Distribution: Brazil; widely cultivated in tropical and subtropical areas as an ornamental, including in the Guianas; 1 collection studied from French Guiana: Route de Baduel, Ile de Cayenne, cultivated, Hoff 5072 (CAY).

Vernacular names: Suriname: bougainville. French Guiana: bougainvillier, bougainville.

Note: A specimen of *Bougainvillea x buttiana* Holttum & Standl. (Publ. Field Mus. Nat. Hist., Bot. Ser. 23: 44. 1944, *B. glabra* Choisy x *B. spectabilis* Willd.), a hybrid with bracts crimson or orange, much crisped, and thus differing from the two species treated above which have bracts essentially purple or magenta (varying to white, pink, salmon or purple), not crisped, was collected in 1945 from a cultivated plant at Mazaruni station, Guyana, det. Sandwith. Apart from the bract characteristics mentioned above, it keys to the parent *B. glabra*, with which it shares a distinctly angled perianth-tube, with antrorsely curved, to 0.5 mm long hairs.

3. **GUAPIRA** Aubl., Hist. Pl. Guiane 1: 308; 3: t. 119, as 'Quapira'. 1775. Type: G. guianensis Aubl.

Torrubia Vell., Fl. Flum. 139. 1829.
Type: T. opposita Vell.

Shrubs or trees; unarmed, glabrous or branchlets often rufous-puberulent. Leaves usually opposite; petiolate; blade entire, often coriaceous. Inflorescences lateral and/or terminal, pedunculate paniculate-corymbose cymes; flowers unisexual, dioecious, small, reddish, yellowish-green, or greenish-white; sessile or pedicellate; exinvolucrate, 2- or 3-bracteolate at base. Male perianth obconic-campanulate, limb 5-dentate, teeth short, induplicate-valvate; stamens 6-10, exserted, filaments short-connate at base, anthers oblong, staminodes present; rudimentary pistil often present. Female perianth tubular, limb narrow, shallowly 5-dentate, in bud rounded or obtuse at apex, rarely with staminodes; ovary elongate-ovoid, sessile, style usually short-exserted; stigma fimbriate or fimbriate-lacerate. Anthocarps drupaceous, red to black, exocarp fleshy, juicy, utricle elongate, papyraceous to coriaceous, striate; seed with hyaline testa adherent to pericarp, embryo straight, cotyledons broad, enclosing scanty endosperm, radicle short, inferior.

Distribution: About 70 species in Central and South America, and the West Indies; 4 species in the Guianas.

KEY TO THE SPECIES
(See also Combined Key to Vegetative Characteristics of Species of *3. Guapira* and *5. Neea* , p. 26)

1 Inflorescence-branches glabrous .... *2. G. eggersiana*
  Inflorescence-branches rufous-puberulent or -tomentellous 2

2 Leaves usually broadly ovate, often cuspidate; female perianth ellipsoid or
  infundibuliform, not constricted at apex, with acute, erect lobes ....
  .... *1. G. cuspidata*
  Leaves usually elliptical or broadly oblong, not cuspidate 3

3 Female perianth densely rufous-puberulent, narrowly tubular ....
  .... *3. G. kanukuensis*
  Female perianth lightly rufous-puberulent, subtubular or ellipsoid ....
  .... *4. G. salicifolia*

1. **Guapira cuspidata** (Heimerl) Lundell, Wrightia 4: 80. 1968. –
   Pisonia cuspidata Heimerl, Bot. Jahrb. Syst. 21: 628. 1896. – *Torrubia cuspidata* (Heimerl) Standl., Contr. U.S. Natl. Herb. 18: 100. 1916.
   Syntypes: Trinidad, Caroni R., Eggers 1413 (male) (B, not seen),
   Eggers 1430 (female) (B, not seen).

Pisonia schomburgkiana Heimerl, Beitr. Syst. Nyctag. 34.1897, nom. nud., fide Heimerl, 1931 scripsit: Type specimen (of inscribed herb. name): Guyana, Ri. Schomburgk 588 (male, coll. 1843) (B, photo!); isotypes, B!, K!. Det. by Heimerl as *Pisonia cuspidata* m.[=mihi] f. *schomburgkiana* m.[=mihi], nom. nud.
Torrubia schomburgkiana Standl., Contr. U.S. Natl. Herb. 18: 101. 1916, nom. nud.
Guapira schomburgkiana Lundell, Wrightia 4: 83. 1968, nom. nud.

Tree 1-18 m, trunk to ca. 20 cm diam.; branches yellowish-brown, rugulose, glabrous; branchlets rufous-puberulent, leafy at apex. Leaves opposite; petiole rufous-puberulent, 1-2.5 cm long; blade thinly coriaceous, usually broadly ovate (sometimes lance-elliptical or ovate-elliptical), 11.5-17.3 x 4.8-6.5 cm, gradually or abruptly acuminate or attenuate to apex, which is usually cuspidate or subcuspidate, with obtuse or acute acumen, subdecurrent at base, glabrous above, densely ferrugineous-puberulent, especially on midvein and principal lateral veins beneath when young, glabrate and dull in age; lateral veins prominent, 6-9 on each side, arcuate, laxly anastomosing near plane margin. Inflorescence a terminal or axillary, densely many-flowered cyme; peduncle stout, 5-7.5 cm long; branches rufous-puberulent or –tomentellous, ascending; ultimate cymules 3-4 to many-flowered, flowers sessile or rarely pedicels 1-3 mm long; bracteoles ovate-triangular, 1 mm long, puberulent. Male perianth oblong-infundibuliform, 3-4 x 1.5-2.5 mm, rufous-puberulent, limb shortly 5-dentate, teeth triangular, obtuse; stamens 7-ca. 9, white, longer ones ca. 8 mm long, shorter ones ca. 6 mm long; rudimentary pistil ca. 3.5 mm long. Female perianth ellipsoid or infundibuliform, 2.5-3 x 1 mm, limb shortly 5-dentate, lobes erect, acute; pistil ca. 2.5-3 mm long, stigma slightly exserted, digitately lacerate-fimbriate. Anthocarp dark red, narrowly ellipsoid, ca. 8 x 3 mm, with purple juice.

D i s t r i b u t i o n : Trinidad, Venezuela, Guyana and Suriname; in Guyana principally found at the edge of forests, in scrub savanna, or at the edges of white sand savanna "islands" along the Courantyne, Cuyuni, Berbice and Rupununi rivers and tributaries, in Suriname on sandy ridges with low forest, and in savanna woods bordering rock pavements; 37 collections studied, of which 21 from the Guianas (GU: 19; SU: 2).

S e l e c t e d   s p e c i m e n s : Guyana: Basin of Rupununi R., Isherton, (male and female), A.C. Smith 2446 (K, NY, U, US); Matope Falls, Cuyuni R., (female), FD 6963 = Fanshawe 3399 (NY); Orealla Savanna, Courantyne R., FD 5387 = Fanshawe 2599 (BRG, K). Suriname: Wia-Wia Bank at Grote Zwiebelzwamp, Lanjouw & Lindeman 1084 (US); Nickerie Distr., area of Kabalebo Dam, Lindeman & Görts *et al.* 458 (US)

Vernacular names: Guyana: bakabakaru (Arawak), mafo, mamudan.

Note: See note on *G. kanukuensis*.

2. **Guapira eggersiana** (Heimerl) Lundell, Wrightia 4: 80. 1968. –
Pisonia eggersiana Heimerl, Bot. Jahrb. Syst. 21: 627. 1896. –
Torrubia eggersiana (Heimerl) Standl., Contr. U.S. Natl. Herb. 18:
100. 1916. Type: Not designated amongst Trinidad and Suriname
specimens cited after original description by Heimerl.   – Fig. 9

Pisonia glabra Heimerl, Bull. Misc. Inf. 1932: 220 (1932). – *Guapira glabra*
(Heimerl) Steyerm., Ann. Missouri Bot. Gard. 74: 617. 1987. Lectotype:
Guyana, Upper Demerara R., Jenman 3978 (K!; isolectotype US!) (here
designated).
Pisonia olfersiana sensu Maguire *et al.* (1948) as to specimens cited [e.g., FD
5074 = Fanshawe 2338 (NY!, U)], (non Link, Klotzsch & Otto, Icon. Pl. Rar.
1: 36, t. 15. 1841).
Pisonia pacurero sensu Pulle (1906) as to specimens cited [Hostmann &
Kappler 947 (K!, NY!), Hostmann 824 (U!)], (non Kunth in Humb., Bonpl.
& Kunth, Nov. Gen. Sp. 2: 218. 1817).
Torrubia olfersiana sensu Standley (1931) as to Guianas specimens cited
[e.g. Ri. Schomburgk 595, 600, 1031], (non (Link, Klotzsch & Otto) Standl.,
Contr. U.S. Natl. Herb. 18: 101. 1916).

Shrub or tree 1.5-14 m, trunk to ca. 20 cm diam. Leaves usually opposite;
petiole to 2 cm long; blade thinly to firmly coriaceous, green, drying dark
brown, elliptical or elliptic-lanceolate, sometimes oblanceolate, 4.5-15 x
2-4(-6.4) cm, apex obtuse, acute or acuminate, base cuneate to rounded,
sometimes somewhat hirtellous at first, glabrous at maturity, sometimes
shining; ca. 5 lateral veins, scarcely prominent. Inflorescence axillary or
terminal, erect, corymbiform or subumbellate, sometimes corymbose-
paniculate, glabrous or nearly so, sometimes sparsely puberulous; peduncle
slender; ultimate branches 0.2-0.5 mm wide, with 3-flowered umbel;
pedicels slender, 3-6 x 0.5 mm; bracteoles 3, lanceolate, ca. 0.5-1 mm, often
ferrugineous-puberulent; flowers puberulous when young, glabrous or
glabrate when mature. Male perianth infundibuliform or infundibuliform-
campanulate, 3.5-4(-6.5) x 2(-3.3) (at widest point near apex) mm, shortly
5-toothed, teeth obtuse, ciliate; stamens 7-8, filaments white, longer ones to
9.5 mm long, to 3 mm exserted, shorter ones ca. 7 mm long, anthers brown,
ellipsoid, 0.5 mm; rudimentary pistil ca. 3.5 mm long, style puberulous.
Female perianth green or white, with pink margin, slightly constricted
above, narrowly ellipsoid, subinfundibuliform or salverform, total length
(2-)3-5 mm, sparsely ferrugineous-puberulous when young, 5-toothed,
teeth, up to 0.5 mm long, ciliate; pistil ca. 3 mm long, style capillary,
distinctly exserted, stigma digitately fimbriate into 15-20 divisions.
Anthocarp red, ellipsoid, 7-10 x 3.8-5 mm, apically coronulate.

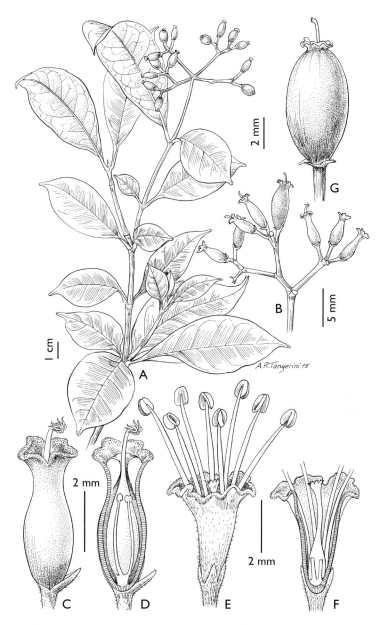

Fig. 9. *Guapira eggersiana* (Heimerl) Lundell: A, habit with female inflorescence; B, female inflorescence; C, female flower; D, section through female flower showing staminodes; E, male flower; F, section through male flower showing rudimentary pistil; G, fruit. (A, C-D, Irwin BG-57; B, E, F, Feuillet 1198; G, Fanshawe 5083). Drawing by A.R. Tangerini.

Distribution: Trinidad, Tobago, northern South America to Brazil; variously in coastal rocky ground, interior mountain escarpments, wallaba (*Eperua falcata*) forests, white sand savannas, and primary and secondary forests in the Guianas; 292 collections studied (3 from Trinidad and Tobago), 283 from the Guianas (GU: 92; SU: 114; FG: 77).

Selected specimens: Guyana: Upper Demerara-Berbice Region, ca. 15 km E of Rockstone, on Linden-Rockstone Road, Pipoly 9601 (U). Suriname: Charlesburg Rift, 3 km N of Paramaribo, Maguire & Stahel 22793 (MO, U). French Guiana: Pic du Grand Croissant, N of Camopi, Feuillet 1198 (U).

Vernacular names: Guyana: hebineroo. Suriname: njamsi-oedoe, njamsihoedoe, jamsi-oedoe , prasara-oedoe, prassa-oedoe, langbladige savanne prasarahoedoe (Sranan); kasoroballi, kassoroballi (Arawak); wayamu sasamuru (Carib). French Guiana: mapou, piment-ramier (Creole); pakaou-meyho, pakaou-kenvi (Oyampi).

Notes: This entity, for which the earliest available name would seem to be *G. eggersiana*, represents the commonest species of *Guapira* in the Guianas. Its name has become involved and entangled with the (still unresolved) identity of *Guapira guianensis* Aubl. (on the plate as '*Quapira guyannensis*'). Woodson & Schery (1961) noted that *G. guianensis* " perhaps is conspecific with *Torrubia eggersiana* (Heimerl) Standl."
The Aublet specimen in the Paris herbarium (P-Jussieu 5170), examined by Howard (J. Arnold Arbor. 64: 270-271. 1983) was said by him to be "a poor match for the plate". We agree with Howard's assessment, as examination of a microfiche of specimen 5170 shows a post-fruiting female stage from which the terminal, 9-flowered inflorescence and shortly campanulate female perianths depicted in Aublet's plate are absent. The specimen's ultimate inflorescence-branches are thin enough to fit our concept of *G. eggersiana*, whereas Aublet's drawing shows thick ones and a stout peduncle, with bracteoles described and drawn as a 5-merous calyx. The Aublet collections at BM do not contain anything relevant to the typification of *G. guianensis* (R. Vickery, pers. comm., 2 Aug. 1995).
The present authors' attribution of the name *G. eggersiana* sensu strictu, which exhibits a (sub)infundibuliform, narrowly ellipsoid or salverform female perianth and characteristically very narrow ultimate inflorescence-branches, to a large proportion of the herbarium material examined as the common Guianan plant, appears to be in accordance with specimens determined by the specialist A. Heimerl as *G. eggersiana* and seen by us. Due to the absence of female flowers on the Aublet specimen, the identity of Guianan plants with campanulate female

corolla must for the time being remain inconclusive.However, a further complication arises in that a significant proportion of those near-*eggersiana* specimens from the Guianas, while not sharing the usual perianth characteristics of that *Guapira* and instead possessing a more or less campanulate female perianth, do indeed possess the thick (more than 0.5 mm wide) ultimate inflorescence-branches characteristic of the Aublet plate. Such plants have, due to the lack of a clear understanding of the identity of *G. guianensis*, traditionally been referred to as *Guapira olfersiana* (Link, Klotzsch & Otto) Lundell in the literature. While the illustration for the basionym of that plant (*Pisonia olfersiana* Link, Klotzsch & Otto, Icon. Pl. Rar. 1: 36, t.15. 1841) indeed shows thick inflorescence-branches and a campanulate female perianth, *P. olfersiana* has meanwhile been sunk by Reitz into the synonymy of the eastern Brazilian *Guapira opposita* (Vellozo) Reitz (Flora Ilustrada Catarinense. Nictaginaceas 32. 1970), in var. *opposita*, which is described as having a tubular ("tubuloso") female perianth. (It may be noted that nyctaginaceous perianths variously described by Reitz as infundibuliform (male *Pisonia aculeata* L. on p. 40) or urceolate-clavate (female *Neea pendulina* Heimerl on p. 25) would be described by the present authors as tubular.)

The array of characteristics shown in herbarium material annotated by various workers as *G. olfersiana* is variable, but in our opinion can probably, at least temporarily, be accommodated under *G. eggersiana* (especially in the absence of biosystematic studies). This position is in accord with, for example, the opinion made at Kew on a plant determined there as *G. (Pisonia) eggersiana*, Fanshawe 5119 (K) from Guyana, which bears the notation: "*Pisonia olfersiana* of Standley [1931] and Maguire [1948]". In conclusion, a situation exists (admittedly unsatisfactory in the absence of a revision) wherein the most cautiously appropriate candidate for the name of (much material of the) commonest species of *Guapira* in the Guianas appears to be *G. eggersiana*, pending future studies of the variability of the plants in the field.

True *Guapira pacurero*, with which *G. eggersiana* has been confused, has leaves drying bright-, pale- or yellowish-green, and occurs in Trinidad, Venezuela and Colombia; it was not reported from the Guianas by Standley (1931).

3. **Guapira kanukuensis** (Standl.) Lundell, Wrightia 4: 81. 1968. – *Torrubia kanukuensis* Standl., Lloydia 2: 178. 1939. Type: Guyana, in dense forest on northwestern slopes of Kanuku Mts. in drainage of Moku-moku Cr. (Takutu tributary), A.C. Smith 3594 (holotype F, not seen; isotypes: (male in young bud) B!; (male in young bud with 2 open flowers) U!). – Fig. 10

40

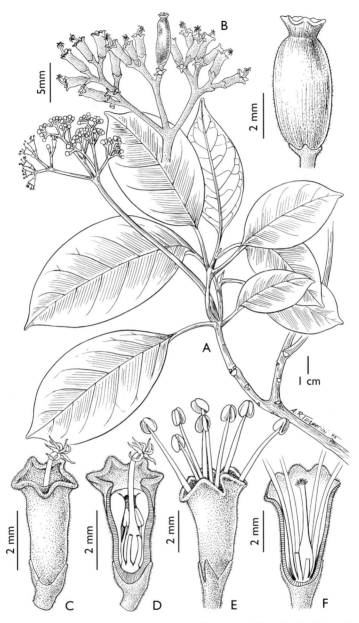

Fig. 10. *Guapira kanukuensis* (Standl.) Lundell: A, habit with male
inflorescence; B, female inflorescence with a fruit; C, female flower; D, section
through female flower showing staminodes; E, male flower; F, section through
male flower showing rudimentary pistil; G, fruit (A-D, G, Smith 3486, E-F,
Smith 3594). Drawing by A.R. Tangerini.

Tree 20-25 m; nodes densely minutely rufous-tomentellous. Leaves opposite or verticillate in whorls of 4; petiole slender, 2-4 cm long, sparsely rufous-puberulous or glabrous; blade membranaceous, elliptical, sometimes oblong-elliptical or ovate, 7.5-18 x 4-8 cm, apex acute, acuminate or obtuse, base often unequal and subrotund to acute, upper surface dark brown when dry, subopaque, glabrous, midvein prominent, lower surface dark brown, glabrous or sparsely minutely puberulous near base; lateral veins ca. 10, smaller veins laxly reticulate. Inflorescence terminal, cymose-paniculate or corymbose, 4-6 x 5-12 cm, with 2-4 ascending or divaricate branches, few- to many-flowered; peduncle to 11.5 cm long, sparsely rufous-puberulent or glabrous; branches densely minutely rufous- or brownish-puberulous; cymules short, few-flowered; bracteoles 4-5, 0.5-1 mm long, densely rufous-tomentellous. Male perianth infundibuliform, 5 x 3 mm, densely rufous-tomentellous, shortly 5-toothed, teeth obtuse; stamens exserted, anthers ca. 0.6 mm long. Female perianth narrowly tubular, 3-4 x 1.5 mm, densely rufous- or brownish-tomentellous, 5-toothed, teeth usually obtuse, flaring or sometimes erect, 0.5 mm long; style exserted to ca. 2 mm, black, stigma fimbriate. Anthocarp narrowly ovoid, ca. 10 x 3.5 mm, densely rufous-tomentellous.

D i s t r i b u t i o n : Endemic to forests of the Kanuku Mts. in Guyana, alt. 150-400 m; 3 collections studied: the type specimen, see above; paratype, A.C. Smith 3486, (F, not seen; isoparatypes B!, K!, NY!, U, US!); Jansen-Jacobs *et al.* 408 (CAY, U).

N o t e : The narrowly tubular, densely rufous-tomentellous female perianth is distinctive for the known Guianan species of this genus. *Guapira marcano-bertii* Steyerm., Ann. Missouri Bot. Gard. 74: 618 (1987) and *G. amacurensis* Steyerm., op. cit. 615 (1987), both described from Delta Amacuro, Venezuela in the Guianan floristic province, may be speculated to possibly occur in Guyana; their relationship to known Guyanan taxa having reddish hairs is yet to be probed. Steyermark (op. cit. 618. 1987) noted that *G. marcano-bertii* has longer peduncles, longer fruits, and longer spreading pubescence throughout, than *G. cuspidata*.

4. **Guapira salicifolia** (Heimerl) Lundell, Wrightia 4: 83. 1968. – *Pisonia salicifolia* Heimerl in Urban, Symb. Ant. 7: 216. 1912. – *Torrubia salicifolia* (Heimerl) Standl., Contr. U.S. Natl. Herb. 18: 101. 1916. Lectotype: Trinidad, Moruga, Broadway 2421 (B, not seen) (designated by Standley, 1931).

Pisonia albiflora Heimerl, Bull. Misc. Inf. 1932: 219. 1932. Type: Guyana, Pomeroon Distr., Kabakaburi, de La Cruz 3317 (holotype K, not seen; isotypes F!, US!).
Torrubia heimerliana Standl., Lloydia 2: 177. 1939. – *Guapira heimerliana* (Standl.) Lundell, Wrightia 4: 81. 1968. Type: Guyana, western extremity of Kanuku Mts., in drainage of Takutu R., A.C. Smith 3146 (holotype F, not seen; isotype U!).

Shrub or tree 1.5-15 m; stem to 20 cm diam.; young branchlets densely rufous-puberulent. Leaves opposite; petiole to 1.5 cm long; blade membranaceous to thinly coriaceous, fuscous when dried, mostly elliptical to elliptic-lanceolate (sometimes oblong-lanceolate or narrowly elliptic-oblong), 5-15 x 1.5-7 cm, subabruptly acuminate or long-acuminate (rarely acute) at apex, cuneate at base, upper surface glabrous, lower surface rufous-puberulent, especially on midvein and lateral veins, glabrate at maturity except usually along midvein and lateral veins; lateral veins 10. Inflorescence terminal or lateral; peduncle to 8 cm long, magenta. Male inflorescence umbelliform; flowers sessile or shortly (to ca. 2 mm) pedicellate; perianth infundibuliform or campanulate, 3-5 x 1.8-2 mm, glabrous to rufous-puberulent, 5-toothed, teeth puberulent, to 1 mm long; rudimentary ovary 2.5-3 mm; stamens (6-)7, anthers 0.7 mm long, light brown. Female inflorescence cymose; bracteoles 2-3, rufous-puberulent, 0.2-1 mm; flowers sessile or shortly (to ca. 2 mm) pedicellate; perianth subtubular, oblong-ellipsoid (or conically tubular), constricted above, (1.5-)2-2.5(-3.5) x ca. 1 mm, including 1-1.5 mm long limb, shortly 5-toothed, yellowish rufous-puberulent; stigma fimbriate-lacerate. Anthocarp ellipsoid-oblong, 8-11.5 x 3-4 mm, thinly and indistinctly striolate, slightly hirtellous, reddish-brown to purplish-black at maturity; seed cream.

Distribution: Trinidad and the Guianas; occurring in primary, hill, gallery and secondary forests, and margins of *Pterocarpus-Euterpe* swamps (Guyana) and *Sesuvium-Avicennia* swamps (Suriname), and elsewhere in French Guiana; 78 collections studied (3 from Trinidad), including 75 from the Guianas (GU: 57; SU: 9; FG: 9).

Selected specimens: Guyana: Mabaruma, Aruka R., North West Distr., (female), FD 5114 = Fanshawe 2378 (NY); Demerara R. (female), Jenman 4870 (NY); Upper Demerara-Berbice Region, Berbice R., 20-30 km SW of Torani Canal, (male), Pipoly 11715 (U), (female) Pipoly 11720 (U). Suriname: Maas & Tawjoeran 10943 (BBS, U); Jonker-Verhoef & Jonker 513 (U). French Guiana: Station des Nouragues, Bassin de l'Arataye, Sabatier & Prévost 1877 (B).

Vernacular names: Guyana: biff wood. Suriname: kleinbladige prasara-oedoe, njamsi-oedoe (Sranan). French Guiana: kumete (Wayana).

N o t e s : Standley (1931) stated that "*Torrubia salicifolia* is probably no more than a form, and not a clearly defined one, of *Torrubia olfersiana*" (note that we place his Guianas specimens of the latter in *Guapira eggersiana*). Perhaps it would indeed seem to be a rufous-puberulent 'version' of *G. eggersiana*. We believe, however, that for the most part the two entities are sufficiently separable to warrant their recognition here, until a revision of the genus in South America is available.

The synonymized *Pisonia albiflora* Heimerl with rufous-puberulent inflorescence-branches resembles *G. eggersiana* in its very narrow inflorescence-branches.

Guapira graciliflora (Mart.) Lundell of Brazil resembles *G. salicifolia* but has a campanulate female perianth. Female flowering specimens (with leaves) of *G. salicifolia* are apt to be superficially very similar to those of female *Neea ovalifolia*, since both taxa have leaves that sometimes dry to yellowish-brown, and both have tubular-cylindrical, densely rufous-puberulent female flowers. The principal recognizable differences between the species are as follows: *G. salicifolia* leaves are not nitid above, and are sparingly rufous-puberulent along the midvein beneath; female flower-buds are obtuse (rounded) at the apex; female flowers are (1.5-) 2-2.5 x 1 mm. In contrast, *N. ovalifolia* has leaves nitid above, glabrous beneath; female flower-buds tapering to long-acuminate at the apex; female flowers are 4 x 1.5 mm.

4. **MIRABILIS** L., Sp. Pl. 177. 1753.
Type: M. jalapa L.

Perennial, sometimes annual herbs, unarmed; taprooted, or with a woody rootstock; stems decumbent or erect, somewhat 4-angled, glabrous or puberulent in lines, occasionally viscid-pubescent, nodes slightly enlarged. Leaves opposite; exstipulate, petiolate or sessile; entire or undulate. Inflorescences axillary or terminal, dense or open, paniculate or thyrsiform cymes; flowers sessile, 1-10 within an involucre; involucre campanulate, calyx-like, 5-lobed, lobes erect, imbricate, equal or unequal, only slightly accrescent, becoming rotate and conspicuously veined in fruit. Flowers bisexual; perianth corolla-like, brightly coloured, tubular, tubular-funnelform, campanulate or rotate, often oblique, deciduous after anthesis, tube slender, often elongate, constricted above ovary, limb 5-lobed, induplicate-plicate, often not broader than tube, lobes retuse or emarginate; stamens 3-5(-6), exserted, filaments capillary, incurved, free or connate at base into a short cup, pollen polyporate; ovary ellipsoid, ovoid, obovoid or subglobose, sessile, ovule anacampylotropous, style filiform, exserted, stigma capitate, papillose. Anthocarp coriaceous, 5-angled, -ribbed or -

costate, more or less constricted at both ends, often tuberculate or rugulose, glabrous or pubescent, mucilaginous when wet; seed with testa adherent to pericarp, embryo uncinate, more or less folded, enclosing mealy endosperm, cotyledons foliaceous, unequal, radicle elongate, descending.

Distribution: Approximately 54 species, mostly in tropical and temperate America, with a center of diversity in the southwestern United States and Mexico, and with 1 species in the western Himalayan region; 1 species in the Guianas.

1. **Mirabilis jalapa** L., Sp. Pl. 177. 1753. Lectotype: Herb. Clifford: 53, Mirabilis No. 1 η (BM, not seen) (designated by LeDuc in Jarvis *et al.* 1993). – Fig. 11

Erect perennial herb to 1.5 m; usually large, fleshy, much-branched taproot. Stem much branched, glabrous or puberulent in lines, rarely villous. Petiole 0.5-5.5 cm long, glabrous or puberulent; blade lanceolate, deltate, ovate-deltate, broadly ovate or lance-ovate, 3.2-14 x 1.8-8.5 cm, apex acute or acuminate, base subcordate or rounded and short-decurrent, glabrous or rarely puberulent, except for ciliate margin. Inflorescence a terminal cyme; peduncle absent or 1-2 mm long; branches ending in 2- to 5-flowered cymose glomerules; involucre 7-19 mm long (longer in fruit), glabrous, puberulent or short-villous, 5-lobed, lobes linear-lanceolate to lance-ovate, ca. 2-3 mm long, acute to attenuate, usually ciliate. Perianth 3-5.5 cm long, variable in colour (pink, red, purple, yellow or white), glabrous or sparsely villous outside, tube 1.5 mm diam., dilated above, limb 2-3.5 cm wide, shallowly lobed; stamens 5, equalling to greatly exceeding perianth. Anthocarp ovoid or obovoid, 7-13 mm long, 5-angled or -ribbed, slightly constricted above base, muricate, verrucose or rugose, glabrous or puberulent, dark brown or black.

Distribution: Tropical America, the exact origin unknown but possibly Mexico (Standley, 1931); pantropically cultivated ornamental, including in gardens of Guyana and Suriname, and occasionally observed in Suriname as probably naturalized; 20 collections studied (GU: 2; SU: 18). The species is cited from Cayenne, French Guiana by Lemée (1955); two specimens at F, from Kamakusa and Demerara in Guyana (both not seen), were cited by Standley (1931).

Selected specimens: Guyana: Pomeroon Distr., Moruka R., Mora Landing, de la Cruz 965 (US). Suriname: Paramaribo, Went 21 (U); Nickerie Distr., Nieuw-Nickerie, probably naturalized, Hekking 951 (B, K, MO, U).

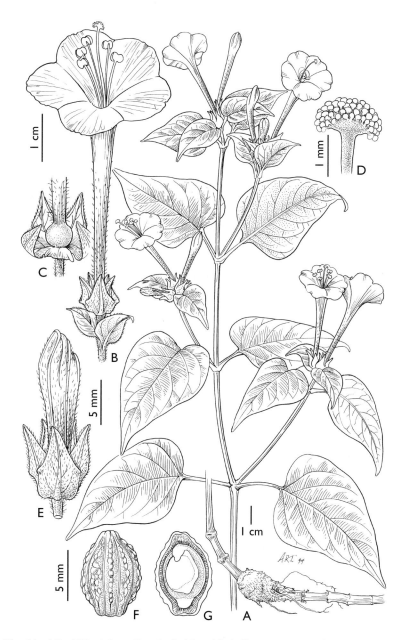

Fig. 11. *Mirabilis jalapa* L.: A, habit with inflorescences and rootstock; B, flower; C, ovary with involucre; D, papillose stigma; E, flower bud with involucre; F, fruit; G, section through fruit (A, Hekking 819; B-G, Smith 1384). Drawing by A.R. Tangerini.

Vernacular names: Guyana: four o'clock, marvel of Peru. Suriname: nachtschone, vieruursbloem (Dutch).

5. NEEA Ruiz & Pav., Fl. Peruv. Prodr. 52. 1794.
Lectotype: N. verticillata Ruiz & Pav.

Trees and shrubs, unarmed; branches erect or subscandent, terete or compressed, glabrous or pubescent. Leaves opposite or in whorls of 4, to irregularly or rarely alternate, sometimes fasciculate; petiolate; blades entire, coriaceous or membranous, glabrous or pubescent. Inflorescences sometimes cauliflorous, axillary or terminal, erect or pendent, pedunculate, dichotomously branched cyme, corymbiform cymes, panicles or thyrses, or flowers sometimes solitary. Flowers unisexual (sometimes bisexual in *N. floribunda*), dioecious, small; sessile or pedicellate, often minutely 1-3-5-bracteolate at base; female flower buds long-acuminate at apex. Male perianth urceolate, globose or elongate, shortly 4- to 5-dentate, teeth induplicate-valvate; stamens 5-10, included, anthers oblong. Female perianth urceolate or tubular-urceolate, constricted above ovary, 4-5-dentate, often contracted at mouth; staminodes often longer than ovary, with distinct, sterile anthers; ovary narrowly ovoid, sessile, included in fleshy base of perianth, style terminal, filiform, stigma exserted, fimbriate. Anthocarps ellipsoid, eglandular, usually crowned by persistent free portion of perianth, often indurate, smooth, striate or costate; seed with hyaline testa adherent to pericarp, embryo straight, cotyledons broad, endosperm scanty, fleshy, radicle short, inferior, horizontal.

Distribution: About 83 species, in South Florida, the West Indies, and tropical America; in the Guianas 5 species.

Note: The genus is named for Luis Nee, a French botanical collector on the Malaspina Expedition (1789-1794) to South and Central America, Mexico, California, the Marianas and Philippines; he earlier became a naturalized Spanish citizen and was associated with the Botanical Gardens in Madrid.

KEY TO THE SPECIES

(See also Combined Key to Vegetative Characteristics of Species of
3. Guapira and *5. Neea*, p. 26)

1 Leaves large, to 25-30(-42) x 12.5-13.8 cm 2
  Leaves often smaller, 7-17(-23) x 2-6.2(-9.5) cm, sometimes verticillate or
  those of a pair sometimes unequal in size 3

2 Tree 4-18 m; leaves glabrous; inflorescence cauliflorous . . . *2. N. floribunda*
  Shrub or small tree to 3 m; leaves rufous-puberulent beneath, densely so on
  veins; inflorescence terminal on small branches, not cauliflorous . . . .
  . . . . *3. N. mollis*

3 Female flowers globose-campanulate, prominently constricted at throat;
  stigma acuminate; leaves drying dark brown . . . . *1. N. constricta*
  Female flowers campanulate, infundibuliform or tubular-cylindrical, not
  constricted; stigma acuminate or fimbriate; leaves drying dark brown or
  greenish- or yellowish-brown 4

4 Leaves often drying greenish- or yellowish-brown, glossy-shining above;
  female flower tubular-cylindrical; infructescence-branches thick and
  subalate . . . . *4. N. ovalifolia*
  Leaves drying dark brown, not glossy-shining above; female flower flared at
  apex; infructescence-branches very thin, not subalate . . . . *5. N. spruceana*

1. **Neea constricta** Spruce ex J.A. Schmidt in Mart., Fl. Bras. 14(2):
368. 1872. Type: Brazil, Pará, near Santarem, Spruce s.n., (U, photo
NY, MG).   – Fig. 12

Shrub or tree 3-12 m, trunk to ca. 15 cm diam., sometimes fluted. Leaves
opposite; petiole to 2 cm long; blade subcoriaceous, elliptical, narrowly
lanceolate or narrowly oblanceolate, 6-17(-21) x 3-6.2(-9.5) cm, often
smaller, those of a pair often markedly unequal in overall size, nearly
subfalcate, acuminate at apex, cuneate at base, glabrous, drying dark
brown. Inflorescence axillary or terminal, divaricately branched, few-
flowered cyme; branches rufous-puberulent; peduncle to 4 cmlong;
bracteoles 4, to ca. 1 mm long, puberulent; flowers subsessile or shortly
pedicellate. Male perianth urceolate, 4 mm long, not constricted, 5-
toothed, pale cream below, with magenta or crimson teeth of ca. 0.2 mm
long, glabrate or minutely and sparsely puberulent; stamens 6. Female
perianth ca. 6 mm long, prominently constricted at throat into a lower
narrow portion of 4 mm long, surrounding ovary, and an upper, globose-
campanulate, ca. 2 mm long limb, glabrate or minutely, sparsely
puberulent, 5-toothed, teeth ca. 0.3 mm long; staminodes ca. 4; style
erect, stigma acuminate. Anthocarp subglobose, 13-17 mm wide, purple.

48

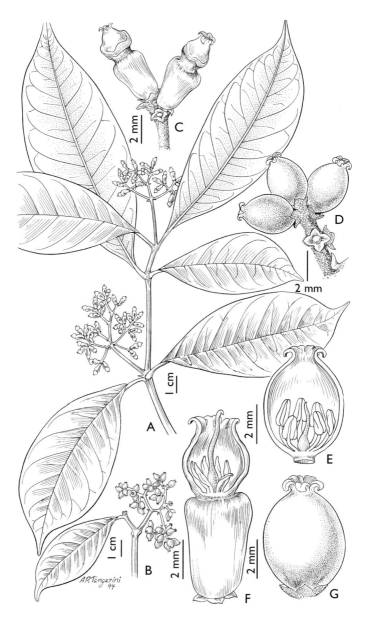

Fig. 12. *Neea constricta* Spruce ex J.A. Schmidt: A, habit with female inflorescences; B, male inflorescence; C, female flowers; D, male flowers; E, section through male flower showing rudimentary pistil; F, female flower, the upper portion sectioned to show staminodes; G, male flower (A-G, Forest Department 4876 and 4171). Drawing by A.R. Tangerini.

Distribution: Brazil, Guyana and French Guiana; in secondary vegetation on white sand in wallaba (*Eperua falcata*) forest in Guyana, and lateritic, open creek forest of stream banks, or non-flooded moist forest in French Guiana; 18 collections studied (GU: 12; FG: 6).

Selected specimens: Guyana: 15 km S of Mabura, along path near Tropenbos Ecological Reserve, (male), Polak *et al.* 424 (U); Mazaruni R., Takutu Cr. to Puruni R., (female), Fanshawe 2140 = FD 4876 (U). French Guiana: Regina Region, eastern plateau of Mts. Tortue, 11 km WNW of Approuague R., Feuillet *et al.* 10188 (U); Saul, vicinity of Eaux Claires, ca. 6 km N of Saül on Route de Belizon, (female), Mori *et al.* 21198 (CAY); Saül, at Eaux Claires, (male), Mori *et al.* 24205 (US).

Notes: The specimens examined include male (Fanshawe 1435 = FD 4171 (NY, K)) and female (Fanshawe 2140 = FD 4876 (NY, U, US)) plants authenticated for Guyana by Maguire *et al.* (1948). The immature fruit with its terminal, constricted, persistent perianth bears a superficial resemblance to the "clove" of commerce, *Myristica fragrans* Houtt. The leaves are very rarely alternate, e.g. Mori *et al.* 23217.
Neea davidsei Steyerm. (Ann. Missouri Bot. Gard. 74: 627. 1987), described from the Delta Amacuro, Venezuela within the Guianan floristic province, may be speculated to possibly occur in Guyana; it is cauliflorous, but with some leaf characteristics apparently similar or comparable to *N. constricta*.

2. **Neea floribunda** Poepp. & Endl., Nov. Gen. Sp. 2: 46. 1835. Type: Peru, San Martin, Mainas, Poeppig 2329 (B, type photo!).

Tree, trunk 4-18 m high, to 40 cm diam., angled and fluted basally; buttresses ca. 40 cm high x 200 cm diam. x 10 cm thick; outer bark whitish or greyish, scaly or smooth. Leaves opposite, or sometimes opposite and alternate on same tree (Mori *et al.* 22686); petiole 1-4 cm long; blade subcoriaceous, glabrous, narrowly to broadly obovate, elliptical or oblanceolate, to 25-30(-42) x (7.5-)13.8 cm, apex obtuse to abruptly acute to shortly or tapering-acuminate, base acute or obtuse, margin sometimes undulate. Inflorescence cauliflorous, or sometimes borne terminally on branches (male flowers), or cauliflorous (female flowers), a many-flowered, lax, dichotomously branched cyme or panicle to ca. 15 x 8 cm; peduncle ca. 3-9.3 cm long, yellow; branches densely and minutely rufous-puberulent. Flowers unisexual, sometimes bisexual; perianth green or yellowish-green when young, rose at base when mature. Male perianth urceolate and generally wider than female perianth, to ca. 5 x 3 mm, sparsely puberulent outside, lobes somewhat ruffled, ca. 0.3-1 x 0.4 mm; stamens 6(9), filaments white, anthers pinkish-white or brown. Female perianth narrowly tubular, 4 x 1.5 mm (including lobes); style with stigma exserted ca. 0.5 mm beyond lobes. Anthocarp ovoid, subgloboid, oblongoid or ellipsoid, to 1.7 x 1.0 mm, glabrous, dark red or reddish-purple.

Distribution: Amazonian Brazil, Peru; in upland rainforest and forest-margins in the Guianas; 108 collections studied (18 from Peru), 90 from the Guianas (GU: 21; SU: 22; FG: 47).

Selected specimens: Guyana: Winiperu Cr., Essequibo R., Fanshawe 291 = FD 3027 (K). Suriname: Bakhuis Mts., between Kabalebo and Left-Coppename Rs., Florschütz & Maas 2857 (U). French Guiana: Saül, at base of Mt. Galbao, Mori et al. 8765 (U).

Use: The Carib Amerindians in Guyana use the fruit for a dye to paint their faces.

Vernacular names: Guyana: humatuba (Warrau). Suriname: njamsi-oedoe, prasara-oedoe (Sranang); marisiballi ojoto (Arawak). French Guiana: milukad kamwi (Palikur); mabechi (Boni); mapou; niamichi oudou (Creole).

Note: Flowers are sometimes bisexual, e.g. Mori et al. 22686 (NY), having 9 well-developed stamens and an ovary with functional ovule (C. Gracie, pers. comm. 1995).

3. **Neea mollis** Spruce ex J.A. Schmidt in Mart., Fl. Bras. 14(2): 367. 1872. Type: Brazil, Amazonas, near S. Gabriel de Cachoeira on the R. Negro, Spruce 2327 (not seen), (U photo, isotype GH).

Subshrubs, shrubs, or small trees 0.8-3(-5) m; branches, leaves and inflorescence-axes densely and softly puberulent with multicellular purple-red, to ca. 1 mm long hairs (hairs brownish or rufous when dry). Leaves opposite or subopposite; petiole to 3 cm long; blade membranaceous, broadly elliptical, to 26.5 x 12.5 cm, obtuse to acuminate at apex, cuneate or attenuate and equal at base, ciliate, glabrous above, rufous-pubescent beneath, hairs densest on veins beneath, becoming glabrate. Inflorescence a terminal or subterminal, few- to many-flowered, branched cyme; flowers subtended by 5 bracteoles of ca. 1 mm long. Male flowers not observed. Female perianth tubular-campanulate, ca. 1-1.5 mm long, constricted just above the ovary, minutely puberulent, green, bluntly 5-dentate, teeth 0.5 mm long. Anthocarp narrowly ellipsoid or oblongoid, to 15 x 8 mm, weakly 4- or 5-angled, minutely puberulent, green when young, deep wine-red when mature.

Distribution: Brazil; in moist forests of Guyana and French Guiana; 51 collections studied, all from the Guianas (GU: 12; FG: 39).

Selected specimens: Guyana: Potaro-Siparuni Region: lower slope of Eagle Mt., McDowell 3489 (U, US); Potaro R., Eagle Mt., Fanshawe 1137 = FD 3873 (K). French Guiana: Saül, Crique Limonade, Maas *et al.* 2238 (U); Saül, Mts. Galbao, Mt. Counana, de Granville 3147 (US).

Use: Employed as a dye plant by the French Guiana Wayapi.

Vernacular names: French Guiana: ka'asala, alasiku (Wayapi).

4. **Neea ovalifolia** Spruce ex J.A. Schmidt in Mart., Fl. Bras. 14(2): 368. 1872. Syntypes: Brazil, Amazonas, near Manáos, Spruce s.n. (U photo Spruce 1855 R. Negro (NY) annotated isotype, cited in Fl. Bras.); Brazil, Pará, Burchell 10011-5 (not seen).

Shrub or tree, 3-21(-30) m x 35 cm, either buttressed and with a crown to ca. 6 m wide, or scrambling; branches glabrous. Leaves opposite or subopposite; petiole to 3 cm long; blade coriaceous, elliptical or oblanceolate, rarely ovate, to 13.5(-23) x 6.5 cm, apex shortly to long-acuminate or acute, base acute to attenuate, glabrous, glossy-shining above, green on both sides, often drying greenish- or yellowish-brown. Inflorescence an axillary or terminal, corymbiform cyme or panicle, all parts rufous-puberulent; peduncle 2-5 cm (male), to 10.5 cm (female) long. Male inflorescence more diffusely branched and flowered than female; bracteoles 5, to 1 mm long; pedicels thin, to 0.4-0.5 mm wide, not subalate. Male perianth urceolate, to 5 x 3 mm, green, sparsely puberulent, 5-dentate or -lobed; stamens 6, anthers ca. 0.8 mm long. Female perianth tubular-cylindrical, 4-5 x 1.5 mm, dirty white; style shortly exserted, fimbriate. Infructescence to 19 cm long; branches subalate and seemingly webbed at insertion area of smaller sub-axes; pedicels thickened to 2 mm diam. Anthocarp ellipsoid, 13-15 x 5.5 mm, ca. 25-sulcate, dark red or black.

Distribution: Brazil; in rainforest and wallaba (*Eperua falcata*) forest in the Guianas, 300-700 m alt.; 82 collections studied, all from the Guianas (GU: 24; SU: 13; FG: 45).

Selected specimens: Guyana: northern slope of Akarai Mts., in drainage of Shodikar Cr. (female), A.C. Smith 2913 (NY, U, US); Guyana-Brazil border, near Kukui Cr. on the Chodikar trail (male), FD

7675 = Guppy 660 (NY). Suriname: Area of Kabalebo Dam project, Nickerie Distr. (female), Lindeman, Görts et al. 623 (U); Lely Mts., Mori & Bolten 8497 (U). French Guiana: Haut Camopi, degrad Belvedere, Prévost & Sabatier 2249 (P).

Vernacular names: Guyana: mamudan; menkaro-kade, menkaroma -kade (Mawayan), koronea, kurunyi (Wai Wai). Suriname: bosgujave; jamsi-oedoe, prasara-oedoe, kleinbladige prasara-oedoe (Sranan). French Guiana: graine-malimbe (Creole); mabechi (Boni).

Notes: A male Guyana specimen examined, Jenman 4840 (BRG, K), previously identified by Heimerl as *Neea nigricans* (Sw.) Heimerl and *Neea jamaicensis* Griseb., is the basis for the inclusion (now considered incorrect) of Guyana in the range of the Jamaican endemic *Neea nigricans* by Heimerl (1934). Similarly, Pulle (1906) cites Splitgerber 1030 (L) from Suriname, as *Pisonia nigricans* Sw. Despite the inference provided by the specific epithet "*ovalifolia*", the leaves of this species are very seldom oval or ovate.
See note under *Guapira salicifolia* regarding similarity to female specimens of *Neea ovalifolia*.

5. **Neea spruceana** Heimerl, Notizbl. Bot. Gart. Berlin-Dahlem 6: 131. 1914. Syntypes: Peru, Dept. Loreto, Tarapoto: Spruce 4858 (not seen); Ule 6498 (male) (B, photo!), 6499 (female).

Shrub or tree 1-16 m; trunk to ca. 20 cm diam; young branches often sparsely rufous-puberulent. Leaves opposite or subopposite, those of a pair often unequal in size, sometimes verticillate in whorls of 3 or 4; subsessile to petiolate, petiole 4-15 mm; blade membranaceous or subcoriaceous, usually elliptical or oblong, sometimes ovate or oblanceolate, 7-16.5(-19.5) x 2-6 cm, apex abruptly or gradually acuminate or long-acuminate, base acute to attenuate, often oblique, glabrous, often drying dark brown. Inflorescence a axillary or terminal, erect, few- to many-flowered, lax, corymbose panicle; peduncle and axes very sparsely and thinly spreading-rufous-puberulent; peduncle 3-6.8 cm long; bracteoles at base of flower 3, triangular-lanceolate, to 0.75-1 mm long; flowers subsessile. Male perianth yellow or yellowish-green, glabrous or with ciliate teeth, urceolate, sometimes infundibuliform or tubular-ellipsoid, 2.4-6(-7.5) x 1-3 mm, 5-toothed, teeth to ca. 2 mm long; stamens 5-6, short ones 2-3.5 mm, long ones 4.5-5 mm long, anthers 0.8-1 mm long. Female perianth green, rose or reddish-purple, minutely rufous-puberulent outside, greenish-white inside, suburceolate, campanulate or infundibuliform, 3 x 1.5-2 mm, limb flared, 5-lobed, lobes to 1 mm long; staminodes 6, ca. 1 mm long; style black, exserted ca. 1.5 mm above perianth, stigma fimbriate into ca. 20 divisions. Anthocarp yellowish-green or maroon-purple, narrowly ellipsoid or oblongoid, ca. 10 x 4 mm, apically crowned by persistent 1 mm long perianth.

Distribution: Colombia, Venezuela, Ecuador, Peru, Bolivia, and Amazonian Brazil; in rainforest, restinga or varzea and in coastal as well as interior forest in Guyana and French Guiana, up to 600 m alt.; 46 collections studied, all from the Guianas (GU: 43; FG: 3).

Selected specimens: Guyana: Puruni, FD 7714 = Boyan 30 (U). French Guiana: Mt. Bellevue de l'Inini, (female), de Granville 7955 (P); Plage de Montjoly, (female), Sauvain 4 (U).

6. **PISONIA** L., Sp. Pl. 1026. 1753.
   Lectotype: P. aculeata L.

Trees and shrubs, sometimes scrambling vines, often armed; young branches pubescent, soon becoming glabrous; spines, when present, axillary and/or produced at tip of a patent branch. Leaves opposite or subopposite; petiolate, entire, glabrous or pubescent. Inflorescences axillary or terminal, sessile or pedunculate, few- to many-flowered, paniculate or corymbiform cymes, sometimes umbelliform or thyrsiform, often borne on modified short shoots. Flowers unisexual, dioecious, exinvolucrate; sessile or pedicellate, 2-3-bracteolate, bracteoles often spirally arranged around the pedicel. Male perianth turbinate or obconic-campanulate, with 5-toothed limb, teeth induplicate-valvate; stamens 6-8(-10), exserted, filaments filiform, connate at base, basifixed or dorsifixed, pollen tricolpate; ovary rudimentary. Female perianth tubular to narrowly campanulate or urceolate, with 5-toothed limb; staminodes present, almost as long as ovary or reduced to a disc; style filiform, slightly longer than perianth, stigma fimbriate, ovule basal, anacampylotropous. Anthocarps dry or fleshy, ellipsoid, oblongoid, obovoid or clavate, with accrescent perianth, terete and costate or 5-angled or -ribbed, with viscid stipitate glands in 1-2(-more) rows along costae, angles or ribs, glands with secretive stalk, head not secretive; seed adherent to pericarp, with deep longitudinal furrow, testa hyaline, embryo straight, cotyledons involute, endosperm scanty, perisperm abundant, mealy, radicle short, inferior.

Distribution: About 40 species in all tropical and subtropical regions, especially America and Southeast Asia.

54

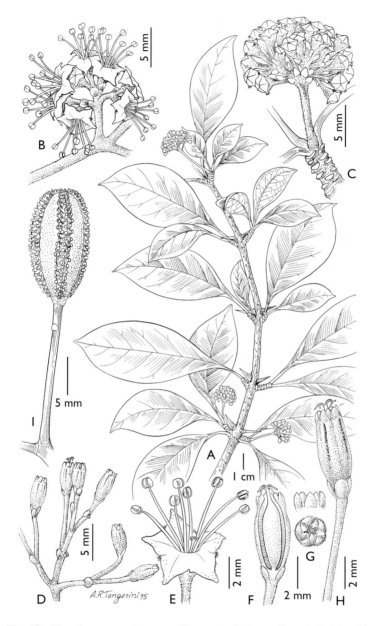

Fig. 13. *Pisonia macranthocarpa* (Donn. Sm.) Donn. Sm.: A, habit with male inflorescences; B, male infloresence; C, male flower buds in inflorescence; D, female inflorescence; E, male flower; F, section through female flower; G, details of perianth-lobes of male flower; H, female flower; I, fruit (A-I, Maas *et al*. 4121, Palmer 197). Drawing by A.R. Tangerini.

1. **Pisonia macranthocarpa** (Donn. Sm.) Donn. Sm., Bot. Gaz. 20: 293. 1895. – *Pisonia aculeata* L. var. *macranthocarpa* Donn. Sm., Bot. Gaz. 16: 198. 1891. Type: Guatemala, Escuintla, Donnell Smith 2091 (US!). – Fig. 13

Scrambling shrub or liana with numerous arcuate branches, climbing to 6 m or more; bark smooth, reddish; stems often armed with axillary, stout, straight to slightly curved, 0.3-1.1 cm long spines, densely pubescent, becoming glabrescent with age. Petiole to 3 cm long; blade subcoriaceous, ovate, elliptical, obovate or suborbicular, 11 x 6 cm, glabrous to sparsely or densely puberulent on upper surface, glabrous or shortly villous on lower surface. Inflorescence seemingly axillary, usually borne at apex of very reduced short shoots, male 1.5-3.5 cm long, female 3.6-10 cm long; peduncle 1-5 cm; bracteoles oblong, 1-1.4 mm long; pedicels short, viscid-pubescent. Male perianth greenish-yellow, broadly campanulate, 2-4 x 2 mm, minutely puberulent-papillate; stamens exserted. Female perianth tubular to more or less urceolate, 1.6-3 mm long; style 1.2 mm. Anthocarp on elongated pedicel, dry, coriaceous, green, obconical or clavate, 16-20 x 8-9 mm, 5-angled and -ribbed, densely puberulent, with 5 rows of stipitate glands, glands in each row often uniseriate, sometimes 2-seriate.

Distribution: Central and South America, including Guyana where occurring in thickets on wooded, elevated "islands" in savanna vegetation; 20 non-Guianan collections studied; 1 collection from a single locality in Guyana, was examined: Rupununi Distr., Chaakoitou, near Mountain Point, just S of Kanuku Mts., (male flowers in bud), Maas *et al.* 4117 (K, U).

30. AIZOACEAE

by

Robert A. DeFilipps and Shirley L. Maina[5]

Predominantly succulent, annual to perennial herbs, subshrubs or shrubs. Leaves opposite, simple; petioles clasping or connate and surrounding stem; stipules present or absent; blades succulent, entire, often pubescent. Inflorescence mostly a terminal dichasium in various forms, sometimes seeming axillary; flowers often solitary by reduction from more complex forms. Flowers bisexual, actinomorphic; tepals coloured inside and green outside, or with petaloid staminodes; perianth-segments (3-)5(-8), free portions often unequal and with dorsal subapical, apiculate appendage, basally connate and adnate to filaments, thus forming a tube; stamens 4-5 or 8-10, arranged in 4-merous whorls or fascicles, if numerous, outer primordia developing into petaloid staminodes, filaments rarely connate at base, anthers dehiscing by longitudinal slits; ovary superior, half-inferior or inferior, (1-)5(-numerous)-carpellate, syncarpous, ovules 1-numerous, anacampylotropous or campylotropous, placentation axile, basal or parietal, styles 1-numerous. Fruit a usually loculicidal, rarely septicidal or circumscissile capsule, sometimes hard and indehiscent, rarely a drupe and occasionally in aggregates; seeds numerous, reniform, mostly exarillate, rarely arillate (aril covering seed completely in *Sesuvium*), embryo curved peripheral, perisperm copious, starchy, endosperm reduced to a layer around radicle.

Distribution: Approximately 2500 species in 127 genera, occurring in the drier parts of New and Old World tropics and subtropics, with a center of frequency in the winter rainfall region of southern Africa; 1 genus and 1 species in the Guianas.

LITERATURE

Bogle, L.A. 1970. The genera of Molluginaceae and Aizoaceae in the southeastern United States. J. Arnold Arbor. 51: 431-462.

Eliasson, U.H. 1996. Aizoaceae. In G. Harling & L. Andersson, Flora of Ecuador 55: 13-27.

Eyma, P.J. 1934. Aizoaceae. In A.A. Pulle, Flora of Suriname 1(1): 158-161.

[5] National Museum of Natural History, Department of Systematic Biology - Botany, NHB 166, Smithsonian Institution, Washington, D.C. 20013-7012, U.S.A.

Hartmann, H.E.K. 1993. Aizoaceae. In K. Kubitzki, The Families and Genera of Vascular Plants 2: 37-69.

Howard, R.A. 1988. Flora of the Lesser Antilles. Aizoaceae. 4: 195-200.

Lemée, A.M.V. 1955-1956. Flore de la Guyane Française. Aizoacées. 1: 580-581. 1955; 4: 33. 1956.

Nevling, L.I. 1961. Aizoaceae. In Woodson, R.E. & R.W. Schery, Flora of Panama 4: 422-427 = Ann. Missouri Bot. Gard. 48: 80-85.

Notes: Regarding the AIZOACEAE, Endress & Bittrich (1993, in same volume as Hartmann, cited above) note that morphologically the family differs from MOLLUGINACEAE by characters in the structure of the androecium and calyx, funicle length and the epidermis of leaves and stems. On a practical basis for distinguishing the families in the Guianas, the following characters can be used:

AIZOACEAE (*Sesuvium*): opposite, succulent cauline leaves, flowers pinkish-purple inside, perianth-segments with an adaxial apical horn, stamens c. 25, and circumscissile capsules containing smooth seeds.

MOLLUGINACEAE (*Mollugo*): whorled, membranaceous cauline leaves, flowers white inside and the segments not apically horned, stamens 3, and valvately dehiscent capsules containing ridged seeds.

Still enigmatic is the identity of a species described as *Mesembryanthemum guianense* by Klotzsch in M.R. Schomburgk, Reisen Brit.-Guiana: 2: 300. 1848, a succulent with dark red flowers collected in the village of Pirara, Guyana in December 1842. Rohrbach in Mart., Fl. Bras. 14(2): 314. 1872 mentioned that he could not find a specimen either at B or K, and apparently there are no later collections of it from the vicinity of Pirara or elsewhere. According to H. Jacobsen (Sukkulentenlexicon: 579. 1970), *M. guianense* is a doubtful species.

1. **SESUVIUM** L., Syst. Nat. ed. 10. 1052. 1759.
Type: S. portulacastrum (L.) L. (Portulaca portulacastrum L.)

Succulent, erect or prostrate, annual or perennial herbs; branches rarely rooting. Leaves fleshy, often more or less terete; exstipulate; petioles sheathing, often connate at base. Inflorescences axillary, of one or more sessile or pedicellate flowers, minutely 2-bracteate at base of pedicel. Flowers perigynous; sessile or pedicellate; perianth-segments 5, imbricate, inside coloured, outside green, appendage dorsal, near apex, long-apiculate; petals absent; stamens either 5, free, or numerous and inserted on perianth-tube, filaments filiform, glabrous; ovary 2-5-locular, ovules numerous, placentation axile, styles (2-)3-5. Fruit a membranous, circumscissile capsule, (2-)3-5-locular, operculum conical, septa incomplete in ripe fruits; seeds numerous in each locule, cochleate to reniform, completely covered by a thin aril, smooth or rough to papillose.

Distribution: Approximately 12, often halophytic, species, occurring pantropically and subtropically; 1 species in the Guianas.

1. **Sesuvium portulacastrum** (L.) L., Syst. Nat. ed. 10. 1058. 1759. – *Portulaca portulacastrum* L., Sp. Pl. 446. 1753. Type: [icon] Hermann, Parad. Bat. t. 212. 1698. – Fig. 14

Sesuvium acutifolium Miq., Tijdschr. Natuurl. Gesch. Physiol. 10: 75. 1843. Type: Suriname, Focke 272 (U).

Perennial herb, glabrous; stem much branched, terete, trailing up to 2 m (often less), often rooting at nodes. Petiole 1-12 mm long; blade often seemingly with salt-exuding hydathode glands above, oblong, elliptical, narrowly oblanceolate, linear-oblong or linear, 1-5(-7) x 0.2-2.5 cm, apex obtuse to acute, base clasping and often overlapping; venation indistinct. Flowers solitary, on a glabrous pedicel to 2 cm long; perianth-tube obconic to subcampanulate, 1.5-3 mm long, lobes lanceolate, ovate-lanceolate or ovate, 4-10 x 2-6 mm, thickened, outside usually prominently veined and green, appendage 1-2 mm long, inside pink, lilac or pinkish-purple, margin scarious pink or hyaline; stamens numerous (ca. 25), unequal, included, inserted at orifice of perianth-tube, filaments 1.5-5 mm long, free above, connate below in a skirt-like tube, anthers reddish, oblong, 0.5-1 x 0.25-0.5 mm; ovary ovoid to subglobose, 3-4 x 2.5-3(-4) mm, styles 3-4, linear, free except at very base. Capsule ovoid, obovoid or conical, 6-11 x 3-6 mm, circumscissile below middle; seeds 8-30, lenticular-reniform or orbicular, black, smooth, shiny, ca.1-1.5 mm wide, aril membranous, transparent.

Distribution: Circumtropical and worldwide subtropical sea shores; often a weed; 45 collections studied, all from the Guianas (GU: 8; SU: 21; FG: 16).

Selected specimens: Guyana: Georgetown, coastal area, Leechman s.n. (K), Hitchcock 16567 (US), Irwin 294 (US). Suriname: Purmerend, mouth of Suriname R., Jonker-Verhoef & Jonker 515 (U); Wia wia-bank on beach ridge, Lanjouw & Lindeman 1066 (U); Coronie, Totness, small coastal ridge, Lanjouw & Lindeman 1522 (U). French

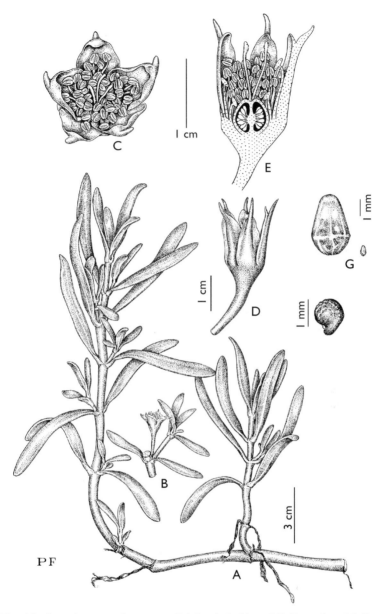

Fig. 14. *Sesuvium portulacastrum* (L.) L.: A, habit, x 2/3; B, node, with flower and branchlet, x 2/3; C, flower, from above, x 4; D. flower, side view, x 2; E, flower longitudinally dissected, x 2; G, fruit, on left, x 6, on right, x 2/3; H, seed, x 10. Drawing by P. Fawcett, reprinted from Correll, D.S. & H.B. Correll. 1982. Flora of the Bahama Archipelago.

Guiana: Battures de Malmanoury, end of Rte. D7, ca. 16 km SE from Sinnamary, Skog & Feuillet 7541 (CAY); Ile de Cayenne, Plage de Zephyr, de Granville 8217 (CAY); Cordon dunaire, NE de Mana, Cremers 8741 (CAY).

Vernacular names: Suriname: strandpostelein (Suriname Dutch). French Guiana: sagu sagu (Carib.).

32. CHENOPODIACEAE

by

ROBERT A. DEFILIPPS AND SHIRLEY L. MAINA[6]

Annual or perennial herbs, rarely shrubs or small trees; monoecious, polygamous or dioecious. Stems sometimes jointed or succulent. Leaves alternate or opposite, simple; exstipulate; blades sometimes much reduced, often farinose, glandular or resin-dotted. Inflorescence of thyrses, or cymes aggregated into spikes, panicles or capitula; bracts present or absent. Flowers bisexual or unisexual, small, green; sessile or shortly pedicellate; perianth absent, or 1-5 tepals, membranous or fleshy, free or basally united; stamens 1-5, filaments mostly free, anthers 2-locular; ovary superior, 1-locular, ovule 1, campylotropous, erect or suspended from a basal funicle, styles and stigmas (1-)2(-3). Fruit a utricle, nut or achene, indehiscent or circumscissile; seed 1, embryo curved, conduplicate, annular or hippocrepiform, endosperm very scant.

Distribution: A worldwide, mostly halophytic family of about 102 genera and 1400 species; 1 genus and 1 species in the Guianas.

LITERATURE

Duke, J.A. 1961. Chenopodiaceae. In R.E. Woodson & R.W. Schery, Flora of Panama 4: 343-348 = Ann. Missouri Bot. Gard. 48: 1-6.

Kellogg, E.A. 1988. Chenopodiaceae. In R.A. Howard, Flora of the Lesser Antilles 4: 139-142.

Kühn, U. 1993. Chenopodiaceae. In K. Kubitzki, The Families and Genera of Vascular Plants 2: 253-281.

Lanjouw, J. 1957. Chenopodiaceae. In A. A. Pulle & J. Lanjouw, Flora of Suriname 1(2): 291-292.

Lemée, A.M.V. 1955. Flore de la Guyane Française, Amarantacées, Chenopodium. 1: 568-569.

Ostendorf, F.W. 1962. Nuttige Planten en Sierplanten in Suriname, Chenopodiaceae. Bull. Landbouwproefstat. Suriname 79: 25.

[6] National Museum of Natural History, Department of Systematic Biology - Botany, NHB 166, Smithsonian Institution, Washington, D.C. 20013-7012, U.S.A.

62

1. **CHENOPODIUM** L., Sp. Pl. 218. 1753.
Lectotype: C. rubrum L.

Annual or perennial, non-succulent herbs, shrubs or small trees; gynomonoecious. Stems glabrous, pubescent or farinose, not jointed. Leaves alternate; at least lowermost usually petiolate; blade foliaceous, entire to pinnatifid, frequently glandular or farinose. Inflorescence of cymes or glomerulate clusters, aggregated into axillary or terminal spikes or panicles, or cymes single and axillary. Flowers bisexual or in part pistillate; ebracteate; tepals (3-)5, free or basally united; stamens (3-)5, alternating and exceeding tepals, filaments flattened, free or basally united, white-hyaline, anthers ovoid, introrse; ovary horizontally flattened, styles and stigmas 2-3. Fruit indehiscent, thin wall adherent or not to seed; seed usually lenticular, shining, black, testa smooth or roughened, embryo annular, hippocrepiform.

Distribution: Approximately 100-150 species in temperate to tropical regions, and widespread as weeds in disturbed places; 1 species in the Guianas.

1. **Chenopodium ambrosioides** L., Sp. Pl. 219. 1753. Type: Herb. Linnaeus No. 313.13 (LINN, not seen). – Fig. 15

Annual or perennial, taprooted herb. Stem 0.3-1.0(-1.5) m tall, strongly scented of mustard, ribbed, often somewhat woody, much-branched. Petiole 1-2 cm long; blade lanceolate, oblanceolate, oblong-elliptic, rhombic-elliptic or ovate, 0.6-12.5 x 1-5.5 cm, entire to shallowly dentate or sinuately pinnatifid, apex acute to obtuse, sometimes apiculate, base cuneate, sessile glandular resin dots, especially on lower surface, glabrous or sparsely puberulent above, or puberulent beneath especially on veins, yellowish-green. Inflorescence a single cyme or spikes of cymes; flowers in glomerules of 4-6, along major axes, in groups of 1-3 on apical, minor axes; glomerules 1-bracteate, bract linear to (sub)foliaceous, to ca. 2.5 cm long; flowers sessile or subsessile. Tepals (3-)5, greenish, narrowly ovate, 0.7-1.3 mm long, glabrous or puberulent and usually gland-dotted, fused ca. half-way, cucullate and folded over fruit; stamens (3-)5; filaments about as long as tepals, anthers orbicular, 0.5 mm long; stigmas sessile or subsessile, spreading. Pericarp not adherent to seed, thin and decaying; seeds lenticular-cochleate to ovoid, 0.6-0.8 mm in diam., horizontally or vertically oriented, smooth, lustrous, reddish-brown.

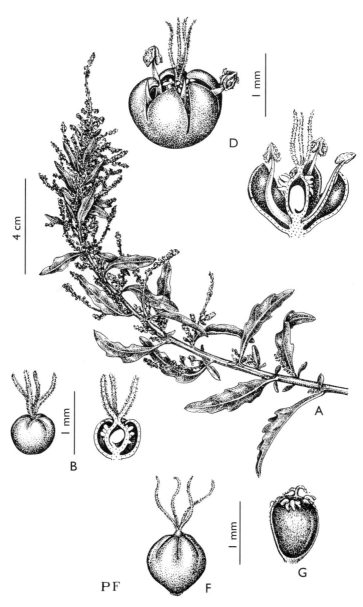

Fig. 15. *Chenopodium ambrosioides* L.: A, flowering branch, x 3/4; B, on left, complete pistillate flower, on right, pistillate flower longitudinally dissected, x 20; D, on left, complete bisexual flower, on right, bisexual flower longitudinally dissected, x 20; F, young fruit, x 20; G, achene, x 20. Drawing by P. Fawcett, reprinted from Correll, D.S. & H.B. Correll. 1982. Flora of the Bahama Archipelago.

Distribution: Possibly native to Mexico and Central America, now a cosmopolitan weed in warm regions; 21 collections examined, all from the Guianas (GU: 6; SU: 1; FG: 14).

Selected specimens: Guyana: South Rupununi Savanna, Aishalton airstrip, Henkel 3467 (US); Rupununi Savanna, Cook 250 (NY, U); Ireng R. near Orinduik Falls, Essequibo County, Irwin *et al.* 474 (US). Suriname: Cultis, Focke 1395 (U). French Guiana: Cayenne, Jardin prima, Kodjoed 91 (CAY); Commune de Remire, Ile de Cayenne, Wittingthon 59 (CAY).

Uses: Generally found as a weed, sometimes cultivated as a medicinal plant for the leaves, which are used as an anthelmintic (vermifuge) in the Guianas (Cook 250; Ostendorf (1962); and Moretti 913). The French Guianans use an infusion of six leaves mixed with salt in a cup, which is reportedly very beneficial for the liver, and as a children's vermifuge (Oldeman B.3909). According to Henkel 3467, the plant is used as a malaria treatment by Wapishiana Amerindians of Guyana.

Vernacular names: Guyana: matouosh; mastruz (Portuguese Guyanese); metroshi (Macushi Amerindian). Suriname: tingi-menti; woron-menti (Creole). French Guiana: woron-wiwiri (Boni); aapoa (Wayapi); zerb' a vers, poudre aux vers (Creole); semen contra.

33. AMARANTHACEAE

by

ROBERT A. DEFILIPPS AND SHIRLEY L. MAINA[7]

Annual or perennial herbs, subshrubs or shrubs, or clambering lianas, or trees. Leaves opposite or alternate, simple, entire or nearly so; sessile to petiolate; exstipulate. Inflorescence of cymules, spikes, racemes, thyrses, heads or panicles, often of ultimate 3-flowered cymules; flowers subtended by 1 bract and 2 bracteoles; lateral flowers of 3-flowered cymules sometimes sterile and modified into scales, spines, bristles or hairs. Flowers bisexual or unisexual, monoecious, dioecious or polygamous when unisexual, regular, hypogynous; tepals (1)3-5 (0 in female *Amaranthus australis*), free or shortly united below, often persistent, scarious, imbricate in bud; stamens (2-3)5, opposite tepals, anthers 2- or 4-locular, dehiscent by longitudinal slits, filaments united below in a cup, pseudostaminodia present or absent; ovary superior, 1-locular, ovules 1-ca. 20 (usually 1), placentation basal, style 1, stigmas 1-3. Fruit a dry, irregularly dehiscent capsule or utricle, rarely berry-like; seeds small, testa shining, embryo horseshoe-shaped or circular, surrounding starchy, copious perisperm, endosperm scanty.

Distribution: A mostly tropical and subtropical family of approximately 1000 species in 69 genera; some weedy or cultivated as ornamental or edible plants; in the Guianas 24 species in 10 genera.

LITERATURE

Burger, W.C. 1983. Flora Costaricensis, Amaranthaceae. Fieldiana, Bot. ser. 2. 13: 142-189.

Cavaco, A. 1962. Les Amaranthaceae de l'Afrique au sud du tropique du Cancer et de Madagascar. Mém. Mus. Natl. Hist. Nat., B, Bot. 13: 1-254.

DeFilipps, R.A. 1992. Ornamental Garden Plants of the Guianas. An Historical Perspective of Selected Garden Plants from Guyana, Surinam and French Guiana. Amarantaceae. pp. 33-35. Smithsonian Institution. Washington, D.C.

Duke, J.A. 1961. Amaranthaceae. In R.E. Woodson and R.W. Schery, Flora of Panama 4: 348-392 = Ann. Missouri Bot. Gard. 48: 6-50.

[7] National Museum of Natural History, Department of Systematic Biology - Botany, NHB 166, Smithsonian Institution, Washington, D.C. 20013-7012, U.S.A.

Eliasson, U.H. 1987. Amaranthaceae. In G. Harling & L. Andersson, Flora of Ecuador 28. 137 pp.

Kellogg, E.A. 1988. Amaranthaceae. In R.A. Howard, Flora of the Lesser Antilles 4: 142-173.

Lachman-White, D.A. *et al.* 1987. A Guide to the Medicinal Plants of Coastal Guyana. Commonwealth Science Council Technical Publication Series 225. 350 pp.

Lemée, A.M.V. 1955-1956. Flore de la Guyane Française, Amarantacées 1: 560-569. 1955; 4(1): 32. 1956.

Nee, M.H. 1995. Amaranthaceae. In J.A. Steyermark *et al.*, Flora of the Venezuelan Guayana 2: 384-399.

Nicolson, D.H. 1991. Flora of Dominica, Part 2. Dicotyledoneae, Amaranthaceae. Smithsonian Contr. Bot. 77: 18-21.

Ostendorf, F.W. 1962. Nuttige Planten en Sierplanten in Suriname, Amaranthaceae. Bull. Landbouwproefstat. Suriname 79: 25-26.

Robertson, K.R. 1981. The genera of Amaranthaceae in the southeastern United States. J. Arnold Arbor. 62: 267-314.

Scheygrond, A. 1932. Amaranthaceae. In A.A. Pulle, Flora of Suriname 1(1): 25-44.

Smith, L.B. & R.J. Downs. 1972. Amarantáceas. In P.R. Reitz, Flora Ilustrada Catarinense. 110 pp.

Townsend, C.C. 1974a. Amaranthaceae. In E. Nasir et S.I. Ali, Flora of West Pakistan 71. 49 pp.

Townsend, C.C. 1974b. Notes on Amaranthaceae - 2. Kew Bull. 29: 461-475.

Townsend, C.C. 1980. Amaranthaceae. In M.D. Dassanayake, A Revised Handbook to the Flora of Ceylon 1: 1-57.

Townsend, C.C. 1993. Amaranthaceae. In K. Kubitzki, The Families and Genera of Vascular Plants 2: 70-91.

Tropicos, Missouri Botanical Garden's nomenclatural database and associated authority files. http://mobot.mobot.org/W3T/Search/vast.html

KEY TO THE GENERA

1 Leaves alternate 2
  Leaves opposite 4

2 Flowers in simple spikes, not glomerulate; tepals more than 4 mm . . . . . .
. . . . 5. *Celosia argentea*
  Flowers in glomerules, these in spikes that may be further branched; tepals less than 4 mm 3

3 Herbs; flowers unisexual; top of ovary rounded or thickened, but not with
  upturned rim; seed exarillate . . . . *3. Amaranthus*
  Lianas (climbing subshrubs); flowers unisexual or bisexual; top of ovary
  with upturned rim; seed arillate . . . . *6. Chamissoa*

4 Plants prostrate, creeping or mat-forming (growing on beaches or in sandy
  places) 5
  Plants erect or only partly decumbent 6

5 Leaves ovate to elliptic or broadly oblanceolate, not succulent; tepals
  glabrous or pubescent; pseudostaminodia present . . . . . *2. Alternanthera*
  Leaves linear to linear-spathulate, succulent; tepals glabrous, except
  ciliolate at base; pseudostaminodia obscure or absent . . . .
  . . . . *4. Blutaparon vermiculare*

6 Inflorescence of 1-3 slender spikes, each generally several dm long, or
  spiciform; flowers reflexed early in development 7
  Inflorescence paniculate, capitate, or if spicate, then at most a few cm long;
  flowers reflexed 8

7 Flowers not in glomerules (except *A. brasiliensis*); bracts white, hyaline;
  bracteoles rigid, subulate, awned, but not hooked . . . .
  . . . . *1. Achyranthes aspera*
  Flowers in few-flowered, shortly pedunculate glomerules, in glomerules
  stiff hooked spines present; bracts and bracteoles modified into hooked
  spines . . . . 7. Cyathula

8 Inflorescence capitate, immediately subtended by a pair of involucral
  leaves; filament-tube extending beyond ovary . . . *8. Gomphrena globosa*
  Inflorescence not capitate or, if so, then the head not immediately subtended
  by a pair of leaves; filament-tube (and pseudostaminodia if present)
  forming cup shorter than ovary 9

9 Inflorescence paniculate 10
  Inflorescence capitate or short-spicate 11

10 Panicle with a diffuse, filigree-like aspect, branches spike-like, flowers not
  in heads, flowers bisexual or unisexual . . . . *9. Iresine diffusa*
  Panicle branches not diffuse, usually (e.g., except in *P. grandiflora*)
  terminating in a 4-7 mm hemispherical head; flowers bisexual . . . . . .
  . . . . 10. Pfaffia

11 Tepals glabrous inside; pseudostaminodia present . . . . *2. Alternanthera*
   Tepals with a short or long tuft of hairs inside; pseudostaminodia absent . .
   . . . . 10. Pfaffia

1. **ACHYRANTHES** L., Sp. Pl. 204. 1753.
Lectotype: A. aspera L.

Annual herbs, sometimes fruticose. Leaves opposite, petiolate. Inflorescence of terminal spikes; bract persistent; bracteoles located on outer side of reflexed flower, becoming indurated with spinose apex. Flowers sessile, becoming reflexed early in development; tepals 5, stiff, persistent; stamens 5, separated by 5 fimbriate pseudostaminodia, filaments and pseudostaminodia joined below in a cup, anthers dorsifixed; style filiform, stigma glandular. Fruit an indehiscent, thin-walled utricle, remaining inside persistent tepals and bracteoles; seed 1, more or less cylindrical, testa thin, membranous, pale brown, embryo closely encircling solid perisperm.

Distribution: Species 6, in the tropics and subtropics of the New and Old Worlds, with a center of diversity in Africa; 1 species in the Guianas.

1. **Achyranthes aspera** L., Sp. Pl. 204. 1753. Type: Ceylon (Sri Lanka), Karalhaebo, Herb. Hermann 2: 69, No. 105 (lectotype BM, not seen) (designated by Townsend 1974a: 35). – Fig. 16

In the Guianas only: var. **aspera**.

Achyranthes indica (L.) Mill., Gard. Dict. ed. 8, *Achyranthes* no. 2. 1768. – *Achyranthes aspera* L. var. *indica* L., Sp. Pl. 204. 1753. Type: not designated.

Herb to 1 m, whitish-pubescent, becoming somewhat woody at base; stems obscurely angled, purplish. Leaf-blade lanceolate, ovate, oblong, elliptical, oblanceolate, suborbicular or orbicular, 0.5-19 x 6 cm, often with undulate margin, apex acuminate or rounded, sometimes apiculate, dark green and pubescent above, lighter green and often more densely pubescent below. Spikes to 60 cm long, slender, with lanate axes, sometimes recurved or flexuous at apex; bracts white, hyaline, narrowly lanceolate, ca. 3 mm long; bracteoles purple-tinged, lanceolate, 3-4 mm long, base membranous and suborbicular, central vein protracted into a tawny, subulate, conspicuously indurated to 1.5 mm long awn. Tepals pinkish, brown, purplish or white, lance-elliptical, 3.7-4.5 mm long, apiculate, with hyaline margin; ovary turbinate, thickened and papillate above, stigma unlobed, scarcely enlarged. Utricle truncate at apex, ca. 2 mm long.

Distribution: This species essentially comprises two elements, of which the Guianan one is a variety originally indigenous to Asia, but now pantropically naturalized as a weed (i.e., var. *aspera*); 22 collections studied, all from the Guianas (GU: 11; SU: 9; FG: 2).

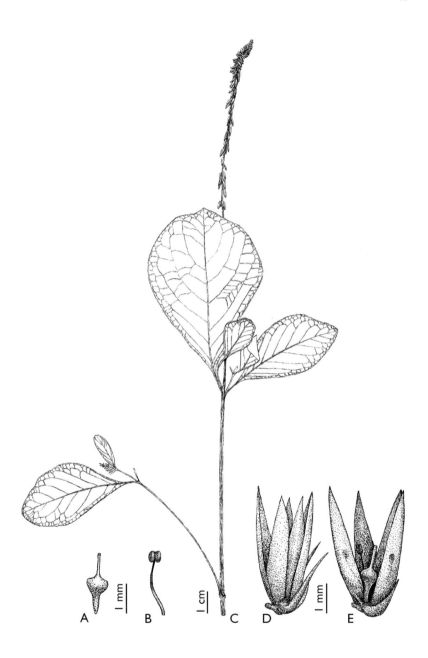

Fig. 16. *Achyranthes aspera* L.: A, fruit; B, stamen; C, habit; D, closed flower; E, opened flower. Drawing by Paul Pardoen.

Selected specimens: Guyana: Near Aishalton Hospital, Stoffers *et al.* 497 (B, CAY, MO, NY, U); Georgetown, Promenade Gardens, Hitchcock 16610 (NY); Mouth of Essequibo R., Forest Bur. Sur., BW 423 (U). Suriname: Plantage Slootwijk, Soeprato 379 (U); Nickerie Distr., Nieuw-Nickerie, Hekking 936 (U); Nickerie, Blufpunt, Teunissen LBB 14975 (U). French Guiana: Iles du Salut, Ile St. Joseph, Debarcadere SW, Cremers 8678 (CAY); Iles du Salut, Ile Royale, Cremers 8457 (CAY).

Uses: In Guyana, the whole plant is decocted and used, either alone or mixed with *Gossypium hirsutum*, in a drink to relieve stomach problems, colds and thrush; a leaf decoction is used to treat high blood pressure; and an infusion of the whole plant is used for heart conditions (Lachman-White *et al.*, 1987).

Note: No Guianan specimens referable to *Achyranthes aspera* L. var. *pubescens* (Moq.) C.C. Towns., a plant indigenous to the New World but differing from var. *aspera* by having tepals 5.2-6.2 mm long, were encountered during this study.

2. **ALTERNANTHERA** Forssk., Fl. Aegypt-Arab. 28. 1775.
Type: A. achyranthes Forssk. [= A. sessilis (L.) DC.]

Telanthera R. Br. in Tuckey, Narr. Exped. Congo 477. 1818.
Type: T. manillensis Walp.

Perennial or annual herbs or subshrubs; young stems and leaves often pubescent, but glabrous with age. Leaves opposite, entire. Inflorescences axillary or terminal, sessile or pedunculate, of various types. Flowers bisexual; tepals 5, outer 3 longer than and enclosing inner 2, glabrous inside; stamens (3)5, united below, anthers oblong or ovate, 2-locular, pseudostaminodia shorter or longer than stamens, subulate or ligulate, distally dentate to fimbriate; ovule 1, style obscure to slender, stigma capitate and globose to more or less punctate. Fruit an indehiscent utricle.

Distribution: Approximately 200 species in the tropics and subtropics of the New and Old Worlds, including 6 species in the Guianas.

KEY TO THE SPECIES

1 Inflorescences of pedunculate heads, at least some peduncles more than 1 cm; largest leaves generally more than 5 cm long 2
Inflorescences of sessile heads, or heads on a peduncle less than 1 cm; largest leaves often less than 5 cm long 3

2 Terrestrial; stems solid, not rooting at nodes; leaves pubescent; bracteoles longer than tepals, with a prominently raised central vein, pubescent on back; peduncles 2-16 cm long; tepals 3.0-4.3 mm long, appressed-puberulent .... *1. A. brasiliana*
Aquatic; stems hollow, often rooting at lower nodes; leaves glabrous or glabrate; bracteoles much shorter than tepals, without prominently raised central vein, glabrous; peduncles 1.0-5.5 cm; tepals 5-6 mm long, glabrous .... *4. A. philoxeroides*

3 Plants always rooting at lower nodes; stems pubescent in 2 lines; bracts and bracteoles less to 0.8 mm long; perianth glabrous; tepals to 2 mm long; stamens usually 3 .... *5. A. sessilis*
Plants sometimes rooting at lower nodes; stems sometimes pubescent in 2 lines; bracts and bracteoles more than 1 mm long; perianth pubescent, at least below; tepals more than 2.5 mm long; stamens 5 4

4 Plants not rooting at nodes; densely velutinous, especially on young stems and undersurface of leaves; largest heads often more than 8 mm long; bracts, bracteoles and tepals erect; tepals 2.6-5 mm long ...... *3. A. halimifolia*
Plants rooting at nodes; glabrescent, velutinous only when young; heads less than 8 mm long; bracts, bracteoles and tepals spreading; tepals 3.7-4.5 mm long .... *2. A. ficoidea*

1. **Alternanthera brasiliana** (L.) Kuntze, Revis. Gen. Pl. 2: 537. 1891. – *Gomphrena brasiliana* L., Cent. Pl. 2: 13. 1756. – *Gomphrena dentata* Moench, Suppl. Meth. 273. 1802, nom. illeg., as 'Gomphraena'. – *Telanthera dentata* Moq. in A. DC., Prodr. 13(2): 378. 1849, nom. illeg. – *Alternanthera dentata* R.E. Fr., Ark. Bot. 16(13): 11. 1920, nom. illeg. Type: [icon] Breyne, Exot. Pl. Cent. t. 52. 1674-1678 (according to Kellogg 1988: 148). – Fig. 17

Perennial herb to 1.5(-2) m, sometimes vining (scandent), scrambling or fruticose; stems erect, spreading, with swollen, purple nodes; stems and young leaves densely appressed-tawny-pubescent. Petiole less than 1 cm long; blade elliptical, lanceolate or ovate, 5-12.5 x 1.5-4.7 cm, green (reddish-purple in cultivar 'Rubiginosa', see note), acuminate at apex, cuneate to rounded at base. Inflorescence axillary or terminal, 2-16 cm

72

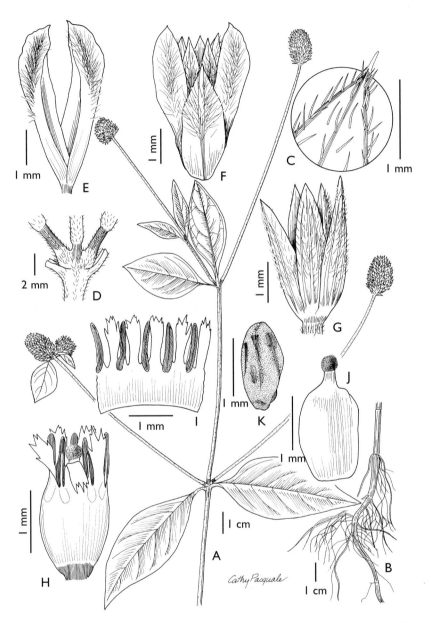

Fig. 17. *Alternanthera brasiliana* (L.) Kuntze: A, habit; B, root system; C, trichomes at leaf apex; D, node; E, bracteoles; F, bracteoles surrounding tepals; G, tepals; H, androecium; I, stamens and pseudostaminodia; J, pistil; K, fruit. (A, Fournet 100; B, ?; C-D, Billiet & Jadin 4518; E-J, Broadway 568). Drawing by Cathy Pasquale.

long pedunculate heads (rarely with a smaller, sessile head in addition to long-pedunculate ones); heads of numerous flowers, hemispherical and to 1 cm long, to conical and to 2.5 cm long, or shortly cylindrical; pedicels minute; bracts ovate to lance-acuminate, mucronate, hyaline, 1-veined, 2.8-4 mm long; bracteoles conduplicate, lanceolate, 4-5.1 mm long, acute, apiculate, serrulate, with stramineous, denticulate margin, pubescent on upper half of back, longer than tepals, with a prominently raised central vein along fold, enclosing tepals. Tepals, white, with hyaline, stramineous margin, narrowly lanceolate, 3.0-4.3 mm long, acute, apiculate, entire, 3-veined, appressed-puberulent on back in a line on the middle; stamens 5, anthers oblong, pseudostaminodia exceeding stamens, ligulate, apically incised or dentate; stigma capitate. Utricle oblong, 2.1 mm long; seeds deep red or brown, cylindrical or subglobose, 1.6-1.8 mm long, shining.

Distribution: New World tropics; in disturbed habitats, roadside weed, or at edges of forest as a vine; 72 collections studied, all from the Guianas (GU: 22; SU: 16; FG: 33).

Selected specimens: Guyana: Basin of Rupununi R., near mouth of Charwair Cr., A.C. Smith 2393 (US); Rupununi Northern Savanna, Moka Moka Cr., S of village, near Kanuku Mts., Goodland 822 (US); Basin of Rupununi R., Isherton, A.C. Smith 2504 (MO). Suriname: Phedra, Middle Suriname R., Kramer & Hekking 2376 (NY); Ningre-Krikie near Moengo road, Lanjouw & Lindeman 535 (NY, U); Upper Suriname R. near Kabelstation along railway at km 130, Jonker-Verhoef & Jonker 605 (U). French Guiana: Region de Cayenne, route montant au Rorota, Garnier 159 (CAY); Cacao, 60 km S de Cayenne, Oldeman B-443 (CAY); Rte de Montjoly, devant l'ancienne usine Prevot, Kodjoed 85 (CAY).

Use: French Guiana: Plant used in a tea for angina (Oldeman BC. 25).

Vernacular names: French Guiana: razier di vin, radie di vin, radie marie-claire (Creole).

Notes: Cultivars with leaves dark reddish-purple, known hitherto as *A.* 'Rubiginosa' and/or 'Ruby', are represented by a specimen grown in a garden in Cayenne, French Guiana (Alexandre 490, CAY).
Alternanthera ramosissima (Mart.) Chodat & Hassl., Bull. Herb. Boiss. ser. 2, 3: 355. 1903, described from Brazil, is recorded for French Guiana by Lemée (1: 567. 1955), probably based on misidentified specimens.

2. **Alternanthera ficoidea** (L.) P. Beauv., Fl. Oware 2: 66. 1818, as 'ficoides'. – *Gomphrena ficoidea* L., Sp. Pl. 225. 1753, nom. cons. Type: Herb. Linnaeus No. 290.23 (typ. cons., LINN).

Alternanthera tenella Colla, Mem. Reale Accad. Sci. Torino 33: 131, t. 9. 1829. Type: Plant cultivated at Torino, Colla s.n. (lectotype P, not seen) (according to Kellogg 1988: 154).
Telanthera bettzickiana Regel, Gartenflora 11: 178. 1862. – *Alternanthera bettzickiana* (Regel) G. Nicholson, Ill. Dict. Gard. 1: 59. 1884, as 'bettzichiana'. – *Alternanthera ficoidea* (L.) P. Beauv. var. *bettzickiana* (Regel) Backer, Fl. Males. Ser. 1. 4: 93. 1949, as 'ficoides'. Type: Plant cultivated at St. Petersburg Bot. Gardens (holotype LE, not seen) (according to Townsend 1974a: 42).
Alternanthera sessilis (L.) DC. var. *amoena* Lem., Ill. Hort. 12: t. 447. 1865. – *Alternanthera ficoidea* (L.) P. Beauv. var. *amoena* (Lem.) L. B. Sm. & Downs, Fl. Ilustr. Catar. 1, Amarantáceas 62. 1972. Type: Plant cultivated at hortus of A. Verschaffelt [Gand] (not seen).

Spreading to ascending woody herb; young stems rooting at nodes, either glabrous or densely pubescent all-round or in 2 lines. Petiole less than 1 cm long; blade ovate to elliptical, 2-9 x 0.5-4 cm, apex acute to acuminate, mucronulate, cuneate at base, glabrous to sparsely pubescent, becoming glabrescent, hairs simple or branched near base. Inflorescence axillary or terminal, sessile heads; heads globose, 3-10 mm wide; bracts and bracteoles white, hyaline, spreading, subequal, ovate to elliptic-lanceolate, 1.5-3.5 mm long, acuminate, aristate, pubescent. Tepals rigid, ovate to elliptical, 3.7-4.5(-5) mm long, acuminate, mucronate or cuspidate, brown to green below, becoming white above, hispidulous, prominently 3-veined, inner 2 tepals navicular; anthers oblong, pseudostaminodia longer than stamens, fimbriate or lacerate apically; style slender, ca. 0.3 mm long, stigma capitate. Utricle ovoid or suborbicular; seed dark reddish-brown, lenticular or cochleate, 1 x 1 mm.

Distribution: West Indies, Panama to South America; common as a ruderal in disturbed areas; 89 collections studied, all from the Guianas (GU: 4; SU: 38; FG: 47).

Selected specimens: Guyana: Demerara, between Demerara and Berbice Rs., de la Cruz 1672 (US); North West Distr., Wanama R., de la Cruz 4030 (US); Pomeroon Distr., Pasanalley Island, de la Cruz 1078 (US). Suriname: Paramaribo, near the Oude Rijweg, Samuels s.n. (US); Paramaribo, Agricultural Experiment Station, Maguire 22715a (NY,US); Saramacca, Experimental Farm Coebiti, Everaarts 794 (CAY). French Guiana: Saül, Mori *et al.* 18780 (CAY, NY); Iles du Salut, Ile Royale, Feuillet 2175 (CAY, U, US); Ile de Cayenne, Colline de Bourda, Le Goff 75 (CAY).

Vernacular name: Suriname: akwemma.

Notes: Since Veldkamp's proposal (Taxon 27: 310-314. 1978) to reject *Alternanthera ficoidea* (L.) P. Beauv. in favour of *A. tenella* Colla, the latter name came in use. At the Berlin Congress (1987), however, the possibility to conserve species names was accepted. The ultimate decision at the Tokyo Congress (1993) was to conserve *Gomphrena ficoidea* with a conserved type. This implies that the recently reintroduced name *A. tenella* should disappear in synonymy, and we can continue with the name that in the Guianas always has been used for this species.

The characteristics of a large species-complex including *A. ficoidea* and *A. halimifolia*, which occurs over much of South America and shows variation in simple vs. branched trichomes, length and apex of bracts and sepals; and inflorescence length, are discussed in detail under *A. tenella* by Kellogg (1988); see also note below under *A. halimifolia*.

The neotropical species *A. paronychioides* A. St.-Hil. is often reported from the Guianas, but no specimens have been encountered in this study. While it may be expected in the Guianas, numerous Guianan specimens of the very similar *A. ficoidea* misidentified as *A. paronychioides* were found. In *A. paronychioides* there are 2 pairs of leaves clustered beneath the inflorescence-head and the bracteoles are ovate, to 2 mm; whereas in *A. ficoidea* there is only one pair of leaves beneath the inflorescence-head and the bracteoles are elliptic-lanceolate, 3.7-4.5 mm long.

Collections of amaranths from the Guianas may sometimes include a cultivated variant of *A. ficoidea*, known as "calico plant", having variegated leaves blotched with red, yellow, orange and/or purple. This *A.* 'Bettzickiana' is grown in the 3 Guianas (GU: Pomeroon Distr., de la Cruz 1078; SU: Paramaribo, Samuels s.n.; FG: Saül, Mori *et al.* 18780). The Bettzickiana plants have leaves narrowly spathulate, and should not be confused with the very similar *A. ficoidea* 'Amoena' having the same colors of variegation as Bettzickiana but with leaves lanceolate or elliptical. The Amoena ("parrotleaf") plant is cultivated and sometimes naturalized in the 3 Guianas (DeFilipps, 1992) [corroborated by SU: Paramaribo, Maguire & Stahel 22715a; FG: Ile Royale, Feuillet 2175], in SU it is known as "Ceylongras" (Ostendorf, 1962).

3. **Alternanthera halimifolia** (Lam.) Standl. ex Pittier, Man. Pl. Usual. Venez. 145. 1926. – *Achyranthes halimifolia* Lam., Encycl. 1: 547. 1785. Type: Peru, Dombey s.n. (holotype P-LAM, not seen; isotype P, not seen) (according to Kellogg 1988: 151).

Telanthera crucis Moq. in A. DC., Prodr. 13(2): 362. 1849. – *Alternanthera crucis* (Moq.) Bold., Fl. Dutch W. Ind. Is. 1: 58. 1909. Syntypes: Caribbean Islands, C. Richard s.n. (G, P, not seen); St. Croix, West s.n. (G-DC, not seen) (according to Kellogg 1988: 149).

Telanthera flavogrisea Urb., Symb. Ant. 1: 300. 1899. – *Alternanthera flavogrisea* (Urb.) Urb., Symb. Ant. 5: 340. 1907. – *Alternanthera tenella* Colla subsp. *flavogrisea* (Urb.) Mears & Veldkamp, Taxon 27: 313. 1978. Type: Jamaica near Rock Fort, Campbell 6059 (holotype B, presumed destroyed) (according to Kellogg 1988: 151).

Woody shrub or herb, to 2 m; stems erect to sprawling and spreading, few-branched; indument on young stems, leaf-veins and inflorescence-axes often yellowish-brown or whitish, densely velutinous with basally branched or stellate hairs, glabrescent with age. Petiole 2-10 mm long; blade mostly narrowly lanceolate or elliptical, sometimes ovate or orbicular, 1-5 x 0.5-2.5 cm, apex obtuse or acuminate, mucronate, cuneate at base. Inflorescence of axillary, sessile, heads; heads ovoid, 4-12 x 3-5 mm; bracts and bracteoles erect, white, hyaline, subequal, broadly ovate to elliptical, 1.4-2.3 mm, acuminate to aristate, pubescent with minutely retrorsely barbed hairs. Tepals erect, dimorphic, outer 3 tepals broader, rigid, ovate to elliptical, 2.6-5.0 mm long, (1-)3-veined, acuminate, mucronate, brown to green below, becoming white above, hispidulous at base with retrorsely barbed hairs, inner 2 tepals shorter, conduplicate, hyaline, sparsely to densely pubescent along fold; stamens 5, anthers oblong, pseudostaminodia exceeding stamens, apically fimbriate or lacerate; stigma capitate. Utricle globose or ovoid, 1.0-1.5 mm; seed reddish-brown, cochleate-orbicular or lenticular, 0.8-1.1 mm wide.

D i s t r i b u t i o n : Circumcaribbean; often in coastal areas; 41 collections studied, all from the Guianas (GU: 27; SU: 11; FG: 3).

S e l e c t e d   s p e c i m e n s : Guyana: Marudi Mts., Mazoa Hill, near Norman Mines camp, Stoffers *et al.* 199 (B, CAY, K, MO, NY, U, US); Kanuku Mts., Rupununi R., Bush Mouth near Witaru Falls, Jansen-Jacobs *et al.* 129 (B, CAY, K, MO, NY, U, US); Rupununi Distr., Chaakoitou, near Mountain Point, just S of Kanuku Mts., Maas & Westra 4119 (B, CAY, MO, NY, U). Suriname: Cultuurtuin, Paramaribo, van Doesburg 57 (U); Plantage Dordrecht, Beneden-Suriname R., Mennega 173 (U). French Guiana: Saül, Skog & Feuillet 731 (B, U).

N o t e : Kellogg (1988: 154-155) suggests this species might be better treated as a variety of *A. ficoidea* (for which the name *A. tenella* still is used), and gives a detailed discussion of its difficult complex, which has four elements occurring in large areas of South America; she includes a table showing the characteristics of certain French Guiana and Suriname variants. The indumentum of this species has been variously characterized as velutinous or sericeous.

4. Alternanthera philoxeroides (Mart.) Griseb., Abh. Königl. Ges. Wiss. Göttingen 24: 36. 1879. – *Bucholzia philoxeroides* Mart., Beitr. Kenntn. Amarantaceen 107. 1825. Type: not designated.

Perennial aquatic herb to 1 m; stems hollow, striate, rooting at lower nodes, pressing to ca. 1.3 cm wide. Petiole 1-16 mm long; blade lanceolate, oblanceolate, oblong-lanceolate or narrowly ovate, 2-10 x 0.5-2.7 cm, entire or sometimes subdenticulate, apex obtuse to acute, base cuneate and tapering to petiole, subcarnose, glabrous above, glabrous or subglabrous beneath. Inflorescence usually axillary (sometimes and/or terminal), 1.0-5.5 cm long pedunculate, solitary heads; heads hemispherical or subglobose, 10-18 x 10-18 mm; bracts and bracteoles subequal, ovate, 1-2 mm long, acuminate, bracteoles much shorter than tepals, glabrous, central vein not prominently raised. Tepals white, oblong-lanceolate, 5-6 x 1.5-2.5 mm, acuminate, mucronulate, denticulate at apex, glabrous, faintly veined; stamens 5, pseudostaminodia exceeding stamens, lacerate; style about twice as long as stigma, stigma capitate. Utricle reniform, 1.0 x 1.0-1.5 mm.

Distribution: Tropical and subtropical South America (adventive elsewhere); weed in aquatic habitats; 23 collections studied, all from the Guianas (GU: 4; SU: 14; FG: 5).

Selected specimens: Guyana: Upper Rupununi R., near Dadanawa, de la Cruz 1515 (MO, NY). Suriname: Paramaribo, way to Kwatta, Samuels 340 (K, NY); Bank of Tibitu R., Lanjouw & Lindeman 1869 (K,NY,U); Para R., 2-4 km S of Houttuinen, Kramer & Hekking 2756 (NY, U). French Guiana: Vicinity of Cayenne, Broadway 371 (NY, US); Cayenne, Sadeeki, Alexandre 380 (CAY).

Vernacular name: Suriname: watra-klaroen.

Note: *Alternanthera maritima* (Mart.) St. Hil. (*A. littoralis* P. Beauv. var. *maritima* (Mart.) Pedersen; *Achyranthes maritima* (Mart.) Standley; *Telanthera maritima* (Mart.) Moq.) is cited for French Guiana (Kourou, Iles du Salut, coll. Sagot) by Lemée (4: 32. 1956, as *Telanthera maritima*), and for "Guiana" by Smith & Downs (1972, p. 56, as *Alternanthera maritima*). It has flowers sessile like *A. philoxeroides*, but differs from the latter by its sessile spikes, whereas *A. philoxeroides* has long pedunculate heads. No Guianan specimens of *A. maritima* were observed in this study.

5. **Alternanthera sessilis** (L.) DC., Cat. Pl. Hort. Monspel. 77. 1813. – *Gomphrena sessilis* L., Sp. Pl. 225. 1753. Type: Ceylon (Sri Lanka), Herb. Hermann 2: 78 (lectotype BM, not seen) (according to Kellogg 1988: 153).

Annual or perennial herb, terrestrial or sometimes shallow-water aquatic, creeping, trailing, scandent or erect, to 60 cm or more, usually smaller; stems ribbed, pubescent in 2 lines in grooves between ribs, or sometimes glabrous, rooting at lower nodes. Leaves of a pair sometimes unequal; petiole short, to 5 mm long; blade ovate, oblanceolate or elliptical, 1-6 x 0.5-1.5 cm, shallowly serrulate, acute to obtuse and mucronulate at apex, cuneate and tapering to base, glabrous on both sides or sparsely puberulent on midvein below. Inflorescence of axillary, congested, sessile, globose heads; heads 2-4 x 3-6 mm; bracts and bracteoles ovate, 0.3-0.8 mm long, acuminate, mucronate, hyaline, prominently 1-veined. Tepals white, hyaline, lance-ovate, narrowly ovate or elliptic-ovate, 1-2 mm long, acute, apiculate, slightly concave, glabrous, 1-veined; stamens 3, shorter than tepals, pseudostaminodia entire, or dentate at apex, shorter than to as long as filaments; style 0.1-0.2 mm long, stigma capitate. Utricle yellowish, bifacially compressed, obcordate, 1.5-2.0 x 1.5-2.2 mm, emarginate, much protruding from and exceeding tepals at maturity, with peg-like persistent style; seed brown or reddish-brown, cochleate-orbicular or lenticular, 1.0-1.2 mm wide, with a sharp margin, shining.

Distribution: Pantropical and subtropical; in disturbed areas; 85 collections studied, all from the Guianas (GU: 20; SU: 37; FG: 29).

Selected specimens: Guyana: Konashen-area, Saparimo, Essequibo R., Jansen-Jacobs *et al.* 1817 (B, CAY, K, MO, U); East Demerara-West Coast Berbice, Abary R. mouth near coastline, Boom 7174 (B, CAY, NY); West Demerara-Essequibo Coast, Pomeroon R., Wakapoa, Bartlett 8024 (NY). Suriname: Saramacca, Experimental Farm Kabo, Everaarts 467 (CAY); Para Distr., Lelydorp vicinity, Santo Swamp, 15 km SW of Paramaribo, Jonker-Verhoef & Jonker 97 (U); Houttuinweg, ca. 7.5 km SE of Paramaribo, Reijenga 28 (U). French Guiana: Sinnamary, Leclerc 15 (CAY); Saül, Mori *et al.* 18456 (CAY, NY); Saint-Laurent du Maroni, ca. 30 km above mouth of Maroni R., St. Laurent-Moute de Maua, Service Forestier 4074.

Note: The very small, glabrous tepals, 3 stamens and obcordate, protruding fruit are characteristic. As exhibited by axes from which fruits have fallen, the globose clusters of flowers are, technically, very densely flowered spikes. Kellogg (1988) noted a phenomenon which tends to be confirmed by our examination of Guianan specimens: "The habit of this species varies considerably depending on the environment. There are two widespread forms of *Alternanthera sessilis*: a graceful, large-leaved (usually) form with large internodal lengths and relatively small inflorescences - in humid to aquatic environs; and a tighter, scrubbier form with smaller leaves, shorter internodes and sometimes very large compactions of axillary inflorescences - in mesic to drier regions."

3. **AMARANTHUS** L., Sp. Pl. 989. 1753.
   Lectotype: A. caudatus L.

   Acnida L., Sp. Pl. 1027. 1753.
   Type: A. cannabina L.

Annual, monoecious or dioecious, herbs; stems erect or prostrate, simple to highly branched, glabrous to pubescent. Leaves alternate, petiolate; blade ovate, elliptical or rhombic, apex usually obtuse or truncate and apiculate, base cuneate and decurrent on petiole; primary and secondary venation often lighter coloured and prominent on upper surface. Inflorescences essentially cymose, glomerules aggregated into axillary clusters, and often also aggregated into terminal spikes, thyrses or panicles; bracts and bracteoles present. Flowers unisexual; tepals 3-5, scarious, apiculate, often persistent; stamens (2)3-5, filaments free, slender, anthers 4-locular; ovary at top rounded or thickened, stigmas 2-3. Fruit a dry, circumscissile or apparently indehiscent utricle; seed lenticular or ovoid, erect, exarillate.

D i s t r i b u t i o n : Approximately 60 species occurring pantropically and in warm temperate regions; 7 species in the Guianas.

L i t e r a t u r e : Costea, M., A. Sanders and G. Waines. 2001. Preliminary results toward a revision of the Amaranthus hybridus species complex (Amaranthaceae). Sida 19: 931-974.

KEY TO THE SPECIES

1 Dioecious; tepals of female flowers absent .... *1. A. australis*
  Monoecious; tepals of female flowers present (3-5 per flower) 2

2 Plants armed with axillary spines .... *6. A. spinosus*
  Plants unarmed 3

3 Leaves strongly emarginate or bilobed at apex . . . . *2. A. blitum*
   Leaves neither strongly emarginate nor bilobed at apex 4

4 Tepals of female flowers 3; stamens 2-3 . . . . *7. A. viridis*
   Tepals of female flowers 5; stamens 5 5

5 Terminal inflorescence 5-20(-25) cm long; tepals of female flowers equal;
   utricle roughly wrinkled or rugulose, at least above . . . . *4. A. dubius*
   Terminal inflorescence often exceeding 30 cm, massive; tepals of female
   flowers unequal, the inner shorter than the outer; utricle smooth or
   rugulose 6

6 Bracts usually 3.5 mm or longer, definitely exceeding the tepals; seeds
   reddish-brown or black . . . . *5. A. hybridus*
   Bracts shorter than 3.5 mm, shorter than the tepals; seeds white, tawny,
   reddish-brown or black . . . . *3. A. caudatus*

1. **Amaranthus australis** (A. Gray) J.D. Sauer, Madroño 13: 15. 1955.
   – *Acnida australis* A. Gray, Amer. Naturalist 10: 489. 1876. Type:
   USA, Florida, A. W. Chapman s.n. (lectotype GH, not seen)
   (designated by Sauer 1955: 15).

Acnida cuspidata Bertero ex Spreng., Syst. Veg. 3: 903. 1826 (non
Amaranthus cuspidatus Vis. 1841). Type: not designated.

Annual, dioecious herb, 1.0-2.5(-4.0) m; stems branched, erect or
decumbent, glabrous, somewhat thickened and hollow at base. Petiole
long, to 10 cm; blade lanceolate, ovate or ovate-lanceolate, 6-30 x 1.5-
14 cm, long-acuminate at apex, acute to rounded at base, undulate,
glabrous. Inflorescence of large, terminal and axillary, lax panicles,
sometimes massed as a flowering top to ca. 30 cm long, often drooping
at apex; bracts 3, ovate, acute, persistent, shorter than tepals. Male
flowers in narrow panicles 3-9 cm long; tepals ovate, 2.5-3 mm long,
mucronate or acuminate, 1-veined; stamens 5. Female flowers in
panicles 6-15 cm long; bracts of female flowers 3, rigid, resembling
tepals; tepals absent; ovary ovoid, subglobose, 3- to 5-angled, stigmas 2-
5. Utricle dehiscent or indehiscent, obovoid or turbinate, 1.5-2.5 mm
long, 3- to 5-angled, as long as or shorter than bracts; seed brown,
lenticular, lustrous, puncticulate.

D i s t r i b u t i o n : Coastal areas of Florida, the Greater Antilles, Mexico,
Trinidad and northern South America; 15 collections studied, 6 from the
Guianas (GU: 1; SU: 3; FG: 2).

Selected specimens: Guyana: Shell Beach, Grewal & Lall 305 (U). Suriname: N. Coronie, Lanjouw 1101 (MO); Wia Wia-bank, Grote Zwiebelzwamp, Lanjouw & Lindeman 1102 (K); Mouth of Corantijne R., Pulle 376 (K). French Guiana: Mana, Savane Sarcelle, de Granville 1971 (CAY); Ilet Zebede, Mana, Hoff 6214 (CAY, U).

Vernacular names: Suriname: zwamp-klaroen, watra-kraroen (Ostendorf, 1962), klaroen (Lanjouw 1101). French Guiana: epinard (Creole) (de Granville 1971).

2. **Amaranthus blitum** L., Sp. Pl. 990. 1753. Type: Europe, Herb. Linnaeus No. 1117.14 (LINN, not seen) (designated by Townsend 1974b: 472).

Amaranthus lividus L., Sp. Pl. 990. 1753. Type: Eighteenth century cultivated specimen (neotype BM, not seen) (designated by Townsend 1974a: 17).

Annual, monoecious, erect to semiprostrate herb to 10 dm; stems subsucculent, glabrous, often pinkish to deep red. Petiole to 5 cm, shorter than to longer than blade; blade often rhomboid, sometimes ovate or elliptical, (0.7-)2-6(-7.5) x (0.2-)2.7-5 cm, usually strongly emarginate, notched or bilobed to 3 mm deep at apex, cuneate at base, glabrous, with prominent veins beneath. Inflorescence of axillary clusters and simple terminal spikes to 9 cm long; bracts linear, acuminate, aristate; bracteoles lanceolate, oblong or ovate, 1/2 to 1/3 as long as tepals, to 1 mm long, acute or obtuse. Male flowers developing far in advance of, and less numerous than female ones; male tepals 3, hyaline, oblong, ovate or elliptical, 0.8-1.3 mm long, acute, mucronate, midvein green, incurved over (2)3 stamens. Female flowers more numerous than male; tepals 3, 2 are linear-oblong to somewhat spathulate, 0.9-2 mm long, about 1/2 to 2/3 the length of mature utricle, obtuse or subacute, third tepal shorter, scale-like; ovary oblong, styles (2-)3, erect. Utricle indehiscent, ovoid-globose, 1.5-2.5 mm long, smooth or somewhat rugulose, thin-walled; seed lenticular, 0.8-1.2(1.6) mm wide, dark reddish-brown or black, shining.

Distribution: Pantropical; weed in disturbed areas; 24 collections studied, all from the Guianas (GU: 5; SU: 15; FG: 4).

Selected specimens: Guyana: Georgetown, Promenade Gardens, Hitchcock 16591 (NY, US); Coast lands, Jenman 5337 (NY, US). Suriname: Lower Saramacca R. near Plantation Catharina Sophia, Lanjouw 278 (K, MO); Lower Corantijne R., Corantijnpolder near

Nieuw Nickerie, Lanjouw 614 (NY); Matappica, near Fisheries Station, Jonker-Verhoef & Jonker 661 (U). French Guiana: Saül, Mori *et al.* 18658 (NY); Route de Baduel, Ile de Cayenne, Hoff 5085 (U); Bourg de Maripasoula, Bassin du Maroni, Fleury 692 (CAY, U).

Vernacular names: Suriname: krakroen, kraroen, klaroen, krassi-wiwiri, redi maka klaroen. French Guiana: mboia (Boni).

Note: Specimens having characteristics which are intermixed, intermediate or intergrading between those of typical *A. blitum* and *A. viridis*, e.g., features of the leaf-apex (bilobed vs. obtuse) and/or utricle surface (smooth vs. rugulose), are frequently encountered, suggesting hybridization and making definite identification to species difficult.

3. **Amaranthus caudatus** L., Sp. Pl. 990. 1753. Type: Herb. Linnaeus No. 1117.26 (lectotype LINN, not seen) (designated by Townsend 1974a: 10).

Annual, monoecious, robust herb to 2 m; stems often red; indument of stems, young leaves and inflorescence-axes pubescent with tangled white hairs. Petiole to 13 cm long, shorter than to as long as blade; blade green to red or red-tinged, lanceolate, elliptical, ovate or rhomboid-ovate, (2.5-)6-20 x (1-)2-8 cm, apex obtuse or acute, often of much smaller size just below inflorescence, glabrous or somewhat puberulent below. Inflorescence of small axillary clusters and massive, lax, tail-like (thus, "*caudatus*") terminal panicles or spikes to 30(-50) cm long; bracts and bracteoles deltoid-ovate, lanceolate or lance-acuminate, 3.5 mm long, shorter than tepals, midvein excurrent, rigid arista at apex. Flowers green, white or reddish, at least some flowers with tepals recurved at apex at maturity. Male flowers interspersed throughout inflorescence; tepals 5, ovate to oblong, 2.5-3.2 mm long, acute; stamens 5. Female flowers more numerous; tepals 5, outer one tepal elliptical or oblanceolate, 1.5-2.8 mm long, acute, inner 4 tepals somewhat shorter than outer one, spathulate or oblong-spathulate, recurved; styles 3, erect. Utricles circumscissile, ovoid-globose, 2.0-2.5 mm, as long as or exceeding perianth, smooth or rugulose; seed subspherical or lenticular, 0.8-1.5 mm, variable in color, white, tawny, reddish-brown or black, dull or lustrous.

Distribution: Origin probably in the Andes Mts. of South America; cultivated as a grain or ornamental and locally escaped throughout the temperate and tropical zones; in the Guianas, the status is usually presumed as an escape from cultivation in former times and now naturalized as a weed; 11 collections examined, 5 from the Guianas (SU: 4; FG: 1).

Selected specimens: Suriname: Kayser Airstrip, 45 km above confluence of Lucie R., Irwin *et al.* 57511 (MO, NY, U); Bakhuis Mts., between Kabalebo and Coppename Rs., Kabalebo Airstrip, Florschütz & Maas 2675 (K, U); Lower Saramacca R. near Catharina Sophia Plantation, Lanjouw 280 (U). French Guiana: Saül, abattis de St. Helene, Gely 74 (CAY).

Vernacular name: Suriname: diega kraroen.

4. **Amaranthus dubius** Mart. ex Thell., Mém. Soc. Sci. Nat. Cherbourg (Fl. Adv. Montpellier) 38: 203. 1912. Type: Plant cultivated at Erlangen Botanic Garden, ex Herb. Schwaegrichen, dedit Hiendlmayr (neotype M, not seen) (designated by Townsend 1974b: 471).

Erect, monoecious, annual herb 2 m, sometimes becoming much branched, stout and succulent; stems green to pink, glabrous to sparsely puberulent. Petiole shorter than to longer than blade; blade ovate, rhombic to lanceolate, 2-9(-17) x 1-6(-11) cm, entire or somewhat crenulate. Inflorescence of axillary clusters and drooping terminal spikes or panicles, 5-20(-25) cm long; bracts and bracteoles lanceolate, ovate or obovate, much shorter than tepals, midrib prominent. Male flowers mostly in apical portion of terminal inflorescence; tepals 5, oblong-elliptical, 1.7-2 mm long, often mucronate; stamens 5(4). Female flowers in axillary clusters and on lower portion of terminal inflorescence; tepals 5, oblong, 1.3-2 mm long, obtuse to acute, often emarginate and mucronate; styles 2-3. Utricle indehiscent to circumscissile, subequal to slightly exceeding perianth, 1.5-2.5 mm long, obtuse or truncate at apex, roughly wrinkled or rugulose, smooth below, often with apparent line of dehiscence; seed cochleate-orbicular or lenticular, 0.7-1.1 mm wide, dark reddish-brown or black, lustrous.

Distribution: Originally tropical America, now pantropical; weed in disturbed areas, or rarely cultivated for the edible leaves in Guyana and French Guiana; 14 collections studied, all from the Guianas (GU: 5; SU: 4; FG: 5).

Selected specimens: Guyana: Coast regions, Jenman 5339 (K); Mahaica, Hitchcock 16771 (NY, US); Georgetown, Promenade Gardens, Hitchcock 16609 (US). Suriname: Paramaribo, Archer 2683 (US); Lelydorpplan, Dirven LP468 (U). French Guiana: Vicinity of Cayenne, Broadway 114 (NY); Cayenne, Jacquemin 2216 (CAY); Ile Royale, Iles du Salut, Cremers 8454 (CAY).

Use: Cultivated for the leaves, which are prepared and eaten as a spinach-like vegetable, in Guyana (Omawale & Persaud 79), Suriname (where sold in Paramaribo markets as greens) (Archer 2677a) and French Guiana (Jacquemin 2216).

Vernacular names: Guyana: calalu, caterpillar calalu, chowri. Suriname: claroen, klaroen. French Guiana: epinard, zergon (Creole).

5. **Amaranthus hybridus** L., Sp. Pl. 990. 1753. Type: Herb. Linnaeus No. 1117.19 (lectotype LINN, not seen) (designated by Townsend 1974a: 11).

Amaranthus paniculatus L., Sp. Pl. ed. 2. 1406. 1763. Type: not designated.

Annual, monoecious (or sometimes flowers bisexual), robust herb to 2.5 m, often smaller; stems green to red; indument of stems, young leaves and inflorescence axes more or less densely pubescent with tangled white hairs or sometimes glabrous. Petiole to 3(-12) cm long; blade subelliptical, ovate, rhombic-ovate or lance-ovate, 2-10(-15) x 1-7(-10) cm, obtuse or acute, slightly notched at apex, entire or minutely crenulate. Inflorescence of large axillary and terminal spikes or panicles 1-20 cm long; bracts lanceolate or ovate, usually 3.5 mm or longer (-4.5 mm), elongated with green midvein extending into a subulate, spinulose apex, conspicuously exceeding tepals so that whole inflorescence often appears bristly. Flowers green, white or reddish. Male flowers occurring in most parts of inflorescence; tepals 5, unequal, narrowly oblong or ovate, 1.3-3.0 mm long, acute; stamens 5. Female flowers, tepals 5, unequal, oblong, longest (outer) 1.3-2.5 mm long, acute, spreading at maturity, shorter than utricle; styles (2)3. Utricle circumscissile, subglobose, 2.0-2.5 mm long, slightly to conspicuously rugulose; seed lenticular or cochleate-orbicular, 0.8-1.1 x 0.9-1.3 mm, dark reddish-brown or black, lustrous.

Distribution: Origin in the New World (eastern North America to northern South America); weed of disturbed land in the Guianas; 14 collections studied, 2 from the Guianas (SU: 1; FG: 1).

Selected specimens: Suriname: Sine loc., Berthoud-Coulon 536 (MO). French Guiana: Ile de Cayenne, Plateau du Mahury, a proximite du lac Rorota, de Granville 5776 (CAY).

6. **Amaranthus spinosus** L., Sp. Pl. 991. 1753. Type: Herb. Linnaeus No. 1117.27 (lectotype LINN, not seen) (designated by Townsend 1974a: 10, as '117.27').

Annual, monoecious or polygamo-monoecious, robust, glabrous herb to 1.5(-2) m; stems slightly woody at base, striate, green, pink or red; rigid stipular spines to 2.5 cm long, axillary, paired, spreading to slightly recurved. Petiole to 6(-8) cm long, about as long as blade; blade ovate to elliptical or rhombic, to 11 x 5 cm, undulate. Inflorescence of axillary clusters, 5-15 mm wide, terminal linear spikes or panicles to ca. 14 cm long; bracts lanceolate to subulate, 1-5 mm long, shorter than or only very slightly exceeding perianth, sometimes subspinose. Flowers green, white or brown, hyaline, midvein green. Male flowers usually at or near apex of terminal portion of inflorescence; tepals 5, lanceolate or oblong, 2.1 x 1.0 mm, acute, apiculate (uter 2 tepals mucronulate); stamens 5, anthers linear, to 2.3 mm long, apiculate. Female flowers more numerous than male flowers; tepals 5, lanceolate, oblong to elliptical, 1.5-2.5 x 1 mm, apex acute to retuse, mucronulate; stigmas 2-3, usually spreading. Utricle irregularly dehiscent or circumscissile, ovoid-conical, smooth or somewhat puckered below; seed lenticular, 0.7-1.1 mm wide, black or (reddish-)brown, lustrous.

D i s t r i b u t i o n : Cosmopolitan, of neotropical origin; often in disturbed lowland habitats; 30 collections studied, all from the Guianas (GU: 5; SU: 14; FG: 11).

S e l e c t e d   s p e c i m e n s : Guyana: Rupununi Distr., Lethem, E bank of Takutu R., Irwin 769 (US); Vreed-en-Hoop, W bank of Demerara R., opposite Georgetown, Hitchcock 16697 (NY, US). Suriname: Saramacca Distr., Experimental Farm Kabo, Everaarts 792 (CAY); Poelepantje, Paramaribo, Jonker-Verhoef & Jonker 243 (U); Near Boskamp at mouth of Coppename R., Wessels Boer 495 (U). French Guiana: Iles du Salut, Ile Royale, Cremers 8429 (CAY, U); Bourg de Saint-Laurent du Maroni, Bassin du Bas-Maroni, Fleury 266 (CAY); Route de Baduel, Ile de Cayenne, Hoff 5085 (CAY).

U s e : In Suriname, the leaves are eaten in salads or (alone) as a vegetable (Ostendorf, 1962; Oldeman 2173).

V e r n a c u l a r   n a m e s : Suriname: maka kraroen, kraroen. French Guiana: epinard, mboya (Boni).

7. **Amaranthus viridis** L., Sp. Pl. ed. 2. 1405. 1763. Type: Herb.
Linnaeus No. 1117.15) (lectotype LINN, not seen) (designated by
Townsend 1974a: 16, as '117.15').

Amaranthus gracilis Desf., Tabl. École Bot. 43. 1804. – *Chenopodium*
caudatum Jacq., Collectanea 2: 325. 1789 (non *Amaranthus caudatus* L.).
Type: not designated.

Annual, monoecious herb to 1 m; stems bluntly triangular, pink to green,
glabrous, or puberulent above. Petiole to 10 cm, not quite as long as blade;
blade ovate-rhombic, ovate, ovate-oblong, lanceolate or bluntly triangular,
(0.3-)1.5 x 6(-9) cm, apex obtuse to truncate, mucronulate, shallowly
emarginate to 1 mm deep, base cuneate to subtruncate, glabrous,
sometimes pilose beneath. Inflorescence of axillary clusters and terminal
spikes or panicles 1-8(-20) cm long; bracts and bracteoles hyaline, ovate,
0.5-1.0 mm long, awned, much shorter than to subequalling perianth.
Flowers white-membranous with green midvein. Male flowers few, near
tips of terminal inflorescences; tepals 3, subequal, elliptical, oblong or
ovate, 1.0-1.5 mm long, concave, acute; stamens 2-3. Female flowers more
numerous than male, in axillary clusters and throughout terminal
inflorescence; tepals 3, oblong, obovate or oblanceolate, 1.2-1.75 mm
long, acute, equalling utricle; ovary oblong, styles 2-3. Utricle indehiscent,
globose or suborbicular, 1.25-1.50 mm long, exceeding perianth,
yellowish-brown, prominently rugulose throughout; seed orbicular,
lenticular, 0.9-1.2 mm long, dark reddish-brown to black, shining.

Distribution: Pantropical; weed on disturbed land, occasionally
cultivated as a leafy vegetable; 14 collections studied, all from the
Guianas (GU: 7; SU: 3; FG: 4).

Selected specimens: Guyana: De Hoop, East Coast Demerara,
Omawale & Persaud 96 (NY); Kamakusa, Upper Mazaruni R., de la
Cruz 4078 (NY, US); Turukwan, Rupununi Savanna, Cook 252 (NY).
Suriname: Upper Suriname R. near Goddo, Stahel 24 (U); Upper
Suriname R. near Saida, Tresling 345 (U); Tabiki, Sauvain 389 (B, CAY,
U). French Guiana: Route de la Chaumiere, Ile de Cayenne, Oldeman B-
2875 (CAY,U); Iles du Salut, Sagot 482, 483 (K).

Use: Leaves are eaten as a vegetable in Guyana (cultivated, Omawale
& Persaud 96) and French Guiana (Oldeman B-2875).

Vernacular names: Guyana: chowroi bhajee. French Guiana:
epinard sauvage (Creole).

Notes: See note concerning specimens of *A. viridis* which may
resemble *A. blitum,* under that species.

Cultivated plants having leaves variegated with blotches of green, red, purple and yellow, and keying in this treatment to the green-leaved *A. viridis* (tepals 3, stamens 3) but with tepals longer than the fruit, are likely referable to *Amaranthus tricolor* L., which is grown as an ornamental in Suriname (since at least the 1930's) and French Guiana, where introduced since the era of Aublet in the late 1700's (DeFilipps, 1992). *Amaranthus polygamus* L., Cent. Pl. 1: 32. 1755 (synonymized under *A. melancholicus* L., Sp. Pl. 989. 1753 by Smith & Downs 1972, p. 27), is a plant of India and Africa, attributed to French Guiana by Lemée (1: 563. 1955) based on a Aublet citation. It has 3 tepals like *A. viridis*, but fruit transversely dehiscent and flowers all in axillary glomerules, whereas *A. viridis* has indehiscent fruit and some flowers in terminal spikes. No specimens of *A. melancholicus* were observed in this study.

4. **BLUTAPARON** Raf., New Fl. 4: 45. 1838.
   Lectotype: B. repens Raf., nom. illeg. (Gomphrena vermicularis L., B. vermiculare (L.) Mears)

Prostrate herbs. Leaves opposite, succulent, very gradually narrowing into a clasping petiole at base, entire, glabrous. Inflorescences of axillary or terminal, sessile or pedunculate heads, or elongated spike-like heads; bracts and bracteoles overlapping, scarious. Flowers bisexual, compressed; tepals 5, thickened and 3-veined from base, outer 3 flat and broad, inner 2 narrower and incurved; stamens 5, filaments free, expanded slightly below, anthers linear, 2-locular, pseudostaminodia absent; ovule 1, long-stalked, style short, stigmas 2, linear. Utricle compressed, ovoid; seed minute.

Distribution: Comprising 4 species in tropical and subtropical regions of the New and Old Worlds; in coastal habitats; 1 species in the Guianas.

Literature: Mears, J. 1982. A summary of Blutaparon Rafinesque including species earlier known as Philoxerus R. Brown. Taxon 31: 111-117.

1. **Blutaparon vermiculare** (L.) Mears, Taxon 31: 113. 1982. – *Gomphrena vermicularis* L., Sp. Pl. 224. 1753. – *Philoxerus vermicularis* (L.) Sm. in Rees, Cycl. 27. 1814. – *Iresine vermicularis* (L.) Moq. in A. DC., Prodr. 13(2): 340. 1849. Type: Hermann s.n., Herb. Sloane 2: 108 (lectotype BM, not seen) (designated by Mears, Taxon 29: 88. 1980; according to Kellogg (1988: 163), however, this choice is disputed by C. Jarvis).   – Fig. 18

88

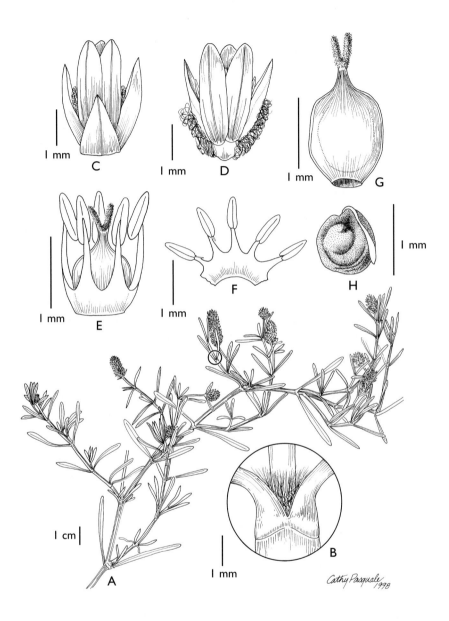

Fig. 18. *Blutaparon vermiculare* (L.) Mears: A, habit; B, detail of node; C, bracteoles subtending tepals; D, tepals; E, androecium and pistil; F, stamens; G, fruit; H, seed. (A-B, Boom 7166; C-H, Pipoly 11249). Drawing by Cathy Pasquale.

In the Guianas only: var. **vermiculare**.

Iresine surinamensis Moq. in A. DC., Prodr. 13(2): 339. 1849. Type: Suriname, Paramaribo, Kappler 1591 (lectotype P, not seen; isolectotype U) (designated by Mears, Taxon 31: 114. 1982).

Perennial herb, diffuse with creeping prostrate branches to ca. 90 cm long, rooting at nodes; stems nodose, pinkish, glabrous except for tufts of hairs in leaf-axils. Petioles of a pair joined and clasping stem at base; blade linear to linear-spathulate, to 6.7 x 0.7 cm, rounded (sometimes acute) at apex, glabrous. Inflorescence a terminal or axillary, solitary, conical, ovoid or cylindrical (rarely clavate) head to 20 x 8 mm, sessile or peduncle to 2.5 cm long; bracts scarious, deltate to lanceolate, 1.8-2.3 mm long, acute; bracteoles 1.2-3 mm long, exceeding bracts but shorter than tepals. Tepals ovate-lanceolate, 3-3.5 mm long, apiculate, scarious, white or pink, minutely ciliolate on sides at base (this hidden by enclosing bases of bracteoles), prominently 3-veined in lower half; filaments included, anthers 0.4 mm long, yellow; ovary included.

Distribution: Variety *vermiculare* occurs throughout the Caribbean region and from Venezuela to Colombia, as well as along sandy coastal areas and mangrove habitats in the 3 Guianas; 65 collections studied, all from the Guianas (GU: 10; SU: 26; FG: 30).

Selected specimens: Guyana: East Coast Demerara, beach at Hope, Robertson & Austin 317 (MO); Abary R. mouth near coastline, Boom 7166 (NY); Foreshore, Georgetown, Demerara, Irwin 289 (US). Suriname: N of Bigie-pan, Lanjouw & Lindeman 3090 (NY,U); Wia Wia Natuurreservaat, Krofajapassie, Pons LBB 12648 (U); Near van Eyk Cr., Geyskes s.n. (U). French Guiana: Iles du Salut, Ile Royale, Cremers 8561 (CAY, U); Sinnamary region, battures de Malmanoury, end of Rte. D7, ca. 16 km SE from Sinnamary, Skog & Feuillet 7542 (CAY, NY); Anse de Montabo, Ile de Cayenne, Hoff 5497 (CAY).

Vernacular names: Suriname: postelein. French Guiana: sagu sagu (Galibi).

Note: This species occurs in the southern United States, West Indies, and from Mexico to South America. Mears (1982) notes that the form represented by the type of *Iresine surinamensis* Moq. (whose characteristics are not mentioned by Mears) is found from the Guianas to Brazil; the U isolectotype is a gnarled, depauperate plant with one capitulum of 9 x 8 mm. The var. *longispicatum* (Moq.) Mears, which occurs in eastern Brazil, is like var. *vermiculare* but differs by having an inflorescence typically 20-28 mm long at maturity; concerning it Mears

(1982) has noted "It is possible that, when more material is available of the Brazilian forms of *B. vermiculare*, the type of *Iresine surinamensis* Moquin might be regarded as properly included with *B. vermiculare* var. *longispicatum*." Additionally, plants ranging from the Grenadines to Venezuela and Colombia having an inflorescence typically of 3 sessile heads, each sometimes more than 7 mm wide, have been separated as var. *aggregatum* (Willd.) Mears. This variety has not been reported from the Guianas, and specimens intermediate between var. *aggregatum* and var. *vermiculare* are said to occur throughout the range of var. *aggregatum*.

5. **CELOSIA** L., Sp. Pl. 205. 1753.
Lectotype: C. argentea L.

Annual or perennial herbs or shrubs, sometimes climbing. Leaves alternate, entire or lobed, subsessile or petiolate; "stipular" leaves sometimes present in axils of normal leaves, lanceolate or falcate. Inflorescence of lax or dense terminal or axillary spikes, thyrses or panicles, in cultivars the spikes sometimes fasciated; bract 1 and bracteoles 2, persistent. Flowers bisexual, not aggregated into glomerules; tepals 5, free, scarious, striate; stamens 5, filaments expanded below and united for less than 1/2 their length in a staminal cup, rarely alternating with short teeth, anthers 4-locular, introrse, pseudostaminodia minute or absent; ovary (1-)2- to multi-ovulate, style short or long, stigmas 2-3. Fruit a circumscissile capsule (sometimes referred to as a utricle), sometimes thickened above; seeds 2 or more, sometimes strongly compressed, orbicular or subovoid, reticulate, shining, exarillate, embryo annular.

Distribution: Approximately 50 species in the world's temperate and tropical zones; 1 species in the Guianas.

1. **Celosia argentea** L., Sp. Pl. 205. 1753. Type: Herb. Linnaeus No. 288.1 (lectotype LINN, not seen) (designated by Townsend 1974a: 5).
– Fig. 19

Celosia cristata L., Sp. Pl. 205. 1753. – *Celosia argentea* L. var. *cristata* (L.) Kuntze, Revis. Gen. Pl. 2: 541. 1891. – *Celosia argentea* L. f. *cristata* (L.) Schinz in Engl. & Prantl, Nat. Pflanzenfam. ed. 2. 16c: 29. 1934. Type: Herb. Linnaeus No. 288.4 (holotype LINN, not seen) (according to Townsend 1974a: 6).

Fig. 19. *Celosia argentea* L.: A, inflorescence; B, flower with tepals removed; C, leaves; D, tepals and style. Drawing by Paul Pardoen.

Annual herb to 1(-2) m, often smaller, sometimes woody at base, glabrous; stems ascending, simple or branched, longitudinally angled. Petiole 0.1-7.0 cm long; blade linear-lanceolate, lance-ovate, lance-elliptic, lanceolate or linear, 2-20 x 0.1-6.0 cm, obtuse to acuminate, sometimes mucronulate, decurrent on petiole. Inflorescence of terminal and upper leaf-axils, densely flowered, up to 12 cm long pedunculate spikes of 2.5-30.0 x 0.7-2.4 cm; bract chaffy, linear, lanceolate or ovate, 2-7 mm long; bracteoles ovate-lanceolate, hyaline, subequal, shorter than bracts; pedicels less than 1 mm long; uppermost flowers occasionally sterile. Tepals lanceolate or elliptic-oblong, 5-11 mm long, acute, scarious-translucent, silvery white to pink, 2- to 5-veined in middle, margin hyaline; stamens cream or pinkish, 3-5 mm long, free filaments longer than fused staminal cup; ovules (1-)3-9, style filiform, 4-8 mm long, stigmas subcapitate, 2- or 3-lobed at apex. Capsule pyriform, ovoid or subglobose, 3-5 mm long; seeds cochleate-orbicular, lenticular, 1.2-2.0 mm wide, black or dark reddish-brown, shining, minutely reticulate or nearly smooth.

Distribution: Tropical Africa, Asia and Pacific islands (possibly originally native to India, fide Robertson, 1981); a wild (non-horticultural) octoploid plant, f. *argentea* is adventive and naturalized as a weed in disturbed tropical and subtropical sites outside its natural range, including in the Guianas; 3 specimens seen, all from the Guianas (GU: 1; SU: 2).

Selected specimens: Guyana: Kamakusa, along upper Mazaruni R., Leng 420 (NY). Suriname: Marowijne R., near Albina, Versteeg 543 (U).

Notes: The tetraploid, fasciated "cockscomb" form of this species, with the apical portion of the inflorescence modified in a bifacially flattened, convoluted, fan-like crest resembling a rooster's comb, is referable to *Celosia* 'Cristata', and individual plants may have flowers with all red, magenta, orange or yellow tepals. It is cultivated as an ornamental in the Guianas, exemplified by Guyana where observed near Turkeyen (DeFilipps, 1992), Suriname where it is called "hanekam" (Ostendorf, 1962); and French Guiana where grown by the Hmongs (Hoff et al. 6281).
Celosia argentea 'Plumosa', the "plume celosia", which bears a colorful diffuse, feathery, plume-like (but non-fasciated) inflorescence, is grown as an ornamental in Suriname (Ostendorf, 1962).

6. **CHAMISSOA** Kunth in Humb., Bonpl. & Kunth, Nov. Gen. Sp. ed. qu. 2: 196. 1818, nom. cons.
Type: C. altissima (Jacq.) Kunth (Achyranthes altissima Jacq.), type cons.

Herbs, shrubs or lianas; stems arcuate. Leaves alternate, membranous, petiolate. Inflorescence of terminal and/or axillary cymes, thyrses or panicles, with spike-like branches bearing few-flowered cymules; bracts 3, stramineous, persistent, much shorter than perianth of lowest flower. Flowers bisexual or functionally unisexual, white or greenish; tepals 5, free, scarious, prominently veined, apiculate; stamens 5, filaments united below into a membranous cup, anthers 4-locular, staminodia absent; ovary at top with upturned rim, stigmas 2(-3). Fruit a dry, 1-seeded, circumscissile capsule; seed with prominent or inconspicuous aril.

Distribution: Species 2, in tropical and subtropical regions of the New World, including the Guianas; mostly occurring in forests and along riverbanks.

Literature: Sohmer, S.H. 1977. A revision of Chamissoa (Amaranthaceae). Bull. Torrey Bot. Club 104: 111-126.

KEY TO THE SPECIES

1 Persistent (fruiting) style 0.5-1.5 mm long; seed with minute, inconspicuous aril . . . . *1. C. acuminata* var. *swansonii*
  Persistent (fruiting) style 0.4-1 mm long; seed usually completely enclosed by loose, membranous aril . . . . *2. C. altissima* var. *altissima*

1. **Chamissoa acuminata** Mart., Beitr. Kenntn. Amarantaceen 78. 1825.
Type: Brazil, Minas Gerais, Martius s.n. (lectotype M, not seen) (designated by Sohmer 1977: 121).

In the Guianas only: var. **swansonii** Sohmer, Bull. Torrey Bot. Club 104: 124. 1977. Type: Panama, Bocas del Toro, Water Valley near Chiriqui Lagoon, Wedel 1695 (holotype GH; isotypes MO, US).

Liana or climber to 1-2 m; stems glabrous or pubescent. Petiole to ca. 5 cm long; blade narrowly ovate to lanceolate, to ca. 15 x 9 cm, apex acuminate, base cuneate to rounded. Inflorescence more or less elongate, often branched; glomerules 1- to 4-flowered; inflorescence-axes puberulent; bracts and bracteoles 1.5-2.5 mm long, pale green, glabrous, keeled, sometimes distally mucronate by prolongation of keel-vein; tepals greenish-white or greenish-yellow, scarious, 2.4-4 mm long, sometimes mucronate, 3-veined; stamens white, shorter than perianth; ovary truncate, with narrow ring of indurated tissue above middle, stigmas 2. Fruit with persistent, 0.5-1.5 mm long style; seed reniform, 1.3-2.5 mm long, aril minute, inconspicuous, testa black or mottled greyish-black, shining, reticulate or minutely foveolate.

Distribution: Mexico through Central America, South America including Colombia, Peru, Paraguay and Suriname; 11 collections studied, 1 from Suriname (SU: 1).

Selected specimen: Suriname: without locality, Hostmann 1109 (U).

2. **Chamissoa altissima** (Jacq.) Kunth in Humb., Bonpl. & Kunth, Nov. Gen. Sp. ed. qu. 2: 197, t. 125. 1818. – *Achyranthes altissima* Jacq., Enum. Syst. Pl. 17. 1760. Type: [icon] Sloane, Voy. Jamaica t. 91, f. 2. 1707 (not seen) (designated by Sohmer 1977: 115). – Fig. 20

In the Guianas only: var. **altissima**.

Climbing liana to 15(-20) m or more; stems glabrous or puberulent. Petiole to 6.5 cm long; blade elliptical, ovate or lanceolate, to ca. 15 x 9 cm, apex acuminate, base cuneate to rounded. Inflorescence more or less elongate, often branched; glomerules 1- to 4-flowered; inflorescence-axes puberulent; bracts and bracteoles 1.5 mm long, pale green, glabrous, keeled, sometimes distally mucronate by prolongation of keel-vein; tepals yellowish-green, whitish-green or cream, scarious, 2.7-4.2 mm long, sometimes mucronate, 3-veined; stamens white, shorter than perianth; ovary truncate, with narrow ring of indurated tissue above middle, stigmas 2. Fruit with persistent, 0.4-1 mm long style; seed reniform, 1.3-2.5 mm long, enveloped in loose, membranous, transparent aril, testa black, shining, minutely striate and finely etched with hexagonal patterns (20 x loupe).

Distribution: Neotropics, from the West Indies and Mexico, through Central America to many parts of South America; 32 collections studied, all from the Guianas (GU: 20; SU: 12).

Selected specimens: Guyana: Mahaica, near canal, Hitchcock 16772 (NY, US); Northwest Distr., above Aracaca, occasional along right bank of Barima R., Cowan 39400 (K, NY); Kanuku Mts., Maipaima,

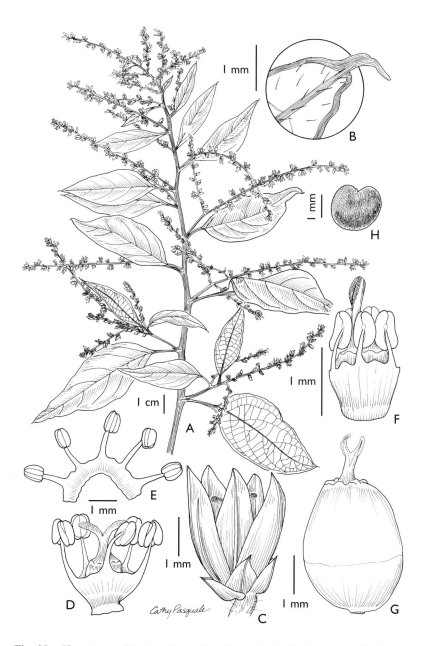

Fig. 20. *Chamissoa altissima* (Jacq.) Kunth: A, habit; B, leaf apex; C, flower with bracteoles; D, flower with tepals removed; E, stamens; F, flower with staminal cup; G, capsule; H, seed. (A-H, Jansen-Jacobs *et al.* 636). Drawing by Cathy Pasquale.

Camp 3 on Tsikoma Cr., Jansen-Jacobs *et al.* 960 (B, CAY, K, MO, NY, U). Suriname: Bank of Tanjimama R., tributary of Coppename R., Mennega 394 (NY, U); Lucie R., 0-2 km below confluence of Oost R., Maguire *et al.* 55423 (NY); Area of Kabalebo Dam project, Nickerie Distr., Lindeman, Görts-van Rijn *et al.* 650 (NY, U).

N o t e : In practice, the difference in length of the persistent style in fruit is the only seemingly reliable characteristic separating the Guianan taxa of *Chamissoa.*

7. **CYATHULA** Blume, Bijdr. 548. 1826, nom. cons.
   Type: C. prostrata (L.) Blume (Achyranthes prostrata L.)

Prostrate to erect, annual or perennial herbs, becoming suffruticose near base. Leaves opposite, entire, petiolate, rhombic-ovate. Inflorescence sometimes capitate, usually comprised of long, terminal spikes and subglobose or cylindrical axillary spikes; fertile and sterile flowers clustered in shortly pedunculate glomerules along major axes; glomerules comprised of bracteate triads of 1 central bisexual fertile and 2 modified sterile flowers formed of uncinate, straight or glochidiate spines, all falling together as fruit; bracts ovate, oblong, acuminate; unisexual flowers reduced to tepals with 5 rigid, hooked awns or spines (glochidia), developing in axils of bracteoles after bisexual flowers. Flowers becoming reflexed early in development; tepals 5, free, often hairy; stamens 5, united at base into a tube, anthers 2-locular, oblong, pseudostaminodia 5, fringed, alternating with fertile stamens; ovary obovoid, ovule 1, style filiform, stigma capitate, glandular. Fruit a thin-walled, indehiscent capsule; seed compressed, ellipsoid.

D i s t r i b u t i o n : Approximately 25-30 species indigenous to the Old and New World tropics; 2 species naturalized as weeds in the Guianas.

KEY TO THE SPECIES

1 Stems erect; glomerules in outline, lanceoloid, acuminate at apex; uncinate
   flowers 2 per glomerule; glochidia of fertile flower nearly twice as long as
   fruiting perianth . . . . *1. C. achyranthoides*
   Stems mostly creeping; glomerules in outline, broadly ovoid or globoid,
   rounded at apex; uncinate flowers 4 per glomerule; glochidia of fertile
   flowers scarcely longer than fruiting perianth . . . . *2. C. prostrata*

1. **Cyathula achyranthoides** (Kunth) Moq. in A. DC., Prodr. 13(2):
326. 1849. – *Desmochaeta achyranthoides* Kunth in Humb., Bonpl. &
Kunth, Nov. Gen. Sp. ed. qu. 2: 210. 1818. Type: Colombia,
Humboldt & Bonpland s.n. (P, not seen) (according to Tropicos).

Erect herb to 1 m, somewhat pubescent, branched; stem striate,
somewhat angled, thickened at nodes. Petiole to ca. 1 cm long; blade
lanceolate, elliptical or ovate, to ca. 11 x 7 cm, acuminate at apex,
tapering at base, pubescent. Inflorescence spiciform, pubescent, with
glomerules distant towards base of axis, congested in middle and distal
parts of axis; glomerules lanceoloid, 3-5 mm long, acuminate, shortly
pedunculate, soon deflexed, usually consisting of 1 fertile, 1 (or 0)
sterile, and 2 rudimentary, unisexual, uncinate flowers, to 2.5 mm long,
nearly twice as long as fruiting perianth. Tepals oblong-lanceolate, acute,
3-veined, pubescent, greenish-white or yellowish; staminodes 3-fid.
Seed brown, smooth.

Distribution: Africa and New World tropics; naturalized as a weed
in disturbed areas; 24 collections studied, all from the Guianas (GU: 9;
SU: 15).

Selected specimens: Guyana: Northwest Distr., Wanama R., de la
Cruz 3997 (MO, NY, US); Tumatumari, Gleason 280 (NY, US); Kanuku
Mts., Nappi-head on Nappi Creek, Camp 1, Jansen-Jacobs *et al.* 588 (B,
CAY, K, MO, NY, U). Suriname: Corantyne R. near Apoera, Jennen &
van der Geest 566 (U); Domburg, lower Suriname R., Plantation "La
Recontre", Jennen & van der Geest 2348 (U); Along Saramacca trail,
above village of Jacob Kondre, Maguire 23879 (NY, U, US).

Vernacular name: Suriname: blaka toriman.

2. **Cyathula prostrata** (L.) Blume, Bijdr. 549. 1825. – *Achyranthes
prostrata* L., Sp. Pl. ed. 2. 296. 1762. Type: Herb. Linnaeus No. 287.13
(lectotype LINN, not seen) (designated by Townsend 1980: 26).
– Fig. 21

Prostrate, weedy herb to 1 m, branched from base; stem striate,
stoloniferous, rooting at lower nodes, thickened at nodes, reddish or
purplish. Petiole 3-18 mm long; blade elliptical to rhombic or obovate-
rhombic, 2-8.5 x 2.5-4 cm, obtuse, narrowed to base and apex, strigose.
Inflorescence 0- to 2-branched from base, spiciform, pubescent, axis
long, narrow, erect, glomerules distant towards middle and base of axis,
to congested near apex; glomerules broadly ovoid or globoid, 1.5-3 mm

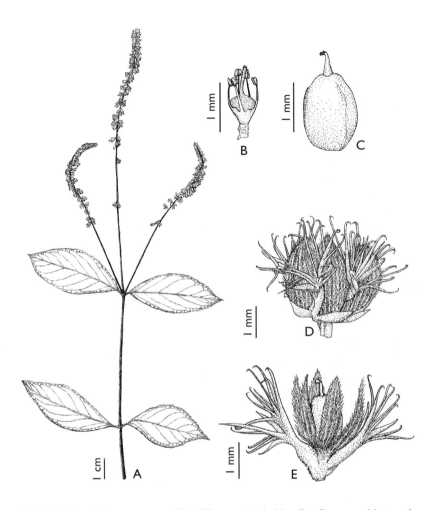

Fig. 21. *Cyathula prostrata* (L.) Blume: A, habit; B, flower with tepals removed; C, fruit; D, bracteate triads of flowers; E, unisexual glochidiate flowers. Drawing by Paul Pardoen.

long, rounded at apex, soon deflexed, usually comprising 2 fertile, 1 sterile and 4 rudimentary, uncinate flowers. Bisexual flowers 2-3 per cluster; tepals scarious, oblong-lanceolate, 1.6-2.4 mm long, mucronate, 3-veined, pubescent. Rudimentary flowers unisexual, greenish-white, about as long as bisexual flowers, reduced to 4-10 uncinate glochidia of ca. 1-1.5 mm long, scarcely longer than fruiting perianth; staminodes 3- or 4-fid. Seed ovoid, smooth, shining, pale brown.

Distribution: Pantropical weed; occurring in recently disturbed or secondary vegetation; 93 collections studied (GU: 39; SU: 7; FG: 47).

Selected specimens: Guyana: Konashen area, Essequibo R., at Saparimo, Jansen-Jacobs *et al.* 1787 (B); Marudi Mts., Mazoa Hill, near Norman Mines camp, Stoffers *et al.* 190 (B); Kanuku Mts., Maipaima, Camp 3 on Tsikoma Cr., Jansen-Jacobs *et al.* 974 (B). Suriname: Plantage Peperpot, Paramaribo, Florschütz 986 (U); Lawa R., near Cottica, Versteeg 271 (U); Taponte, Rombouts 763 (U). French Guiana: Saül, Mori *et al.* 18261 (NY); Cacao, env. 60 km S de Cayenne, Oldeman B-466 (NY); Haut Oyapock, les Trois Sauts, Oldeman B-3265 (CAY).

Vernacular names: French Guiana: hamac de la biche (translation of Wayapi); bululu, kinikini (Boni); chaine d'enfant (Creole); temeku (Saramaccan).

Note: The fully spread-out, ovoid shape of the glomerule is due to its having more flowers per glomerule than in *C. achyranthoides*.

8. **GOMPHRENA** L., Sp. Pl. 224. 1753.
   Lectotype: G. globosa L.

Pubescent herbs. Leaves opposite, sessile or shortly petiolate, acute to acuminate, apiculate. Inflorescence axillary or terminal, cylindrical or subglobose, subtended by a pair of involucral leaves; bracts scarious, somewhat cucullate, keeled, keel usually excurrent; bracteoles scarious, conduplicate, winged or crested along fold, enclosing flowers, falling with fruits. Flowers bisexual; tepals 5, scarious, concave, densely lanate on outer surface; staminal tube long, anthers 2-locular, pseudostaminodia absent; stigmas 2, linear, erect, somewhat elongated. Utricle membranous, compressed, ovoid, indehiscent; seed 1.

Distribution: Approximately 95 species indigenous to the New World tropics; 1 species naturalized and cultivated in the Guianas.

Literature: Holzhammer, E. 1955-1956. Die amerikanischen Arten der Gattung Gomphrena L. Mitt. Bot. Staatssamml. München 2: 85-114; 178-257.

1. **Gomphrena globosa** L., Sp. Pl. 224. 1753. Type: Herb. Linnaeus No. 319.1 (lectotype LINN, not seen) (designated by Townsend 1974a: 46).   – Fig. 22

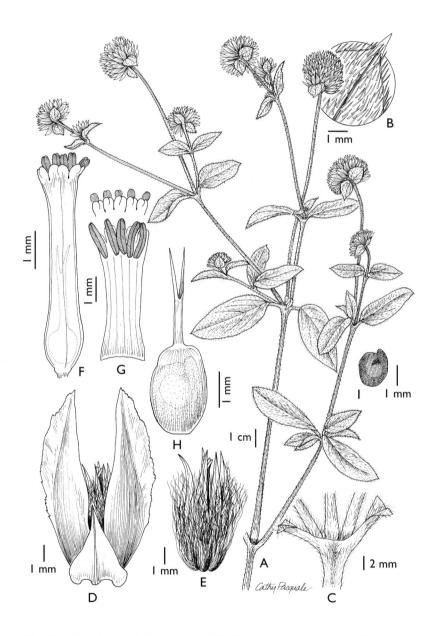

Fig. 22. *Gomphrena globosa* L.: A, habit; B, trichomes at leaf apex; C, node; D, bracts and bracteoles; E, lanate tepals; F, androecium; G, staminal tube opened; H, fruit; I, seed. (A-I, de la Cruz 3579). Drawing by Cathy Pasquale.

Annual herb to 1 m or more; stem pubescent. Leaf-blade oblong, ovate or elliptical, to 11 x ca. 3.5 cm, apex obtuse, acute or acuminate, apiculate, base decurrent onto a short petiole, entire, pubescent, ciliate. Inflorescence of terminal and axillary, to ca. 15 cm long pedunculate, hemispherical or short-cylindrical heads; heads to 2.5 x ca. 2.5 cm; involucral leaves green, foliaceous, broadly ovate, to ca. 1 cm long, pubescent like cauline leaves; bracts ovate, 4-6 mm long, scarious, glabrous; bracteoles narrowly elliptical, 7-10 mm long, glabrous, strongly scarious-winged, irregularly jagged-denticulate, much longer than perianth. Flowers red, white, purplish-red, pink or yellow; tepals 5-6 mm, lanate with straw-colored, whitish, reddish or brownish hairs. Utricle ellipsoid, strongly apiculate into a to ca. 2 mm long beak; seed lenticular-ovoid, ca. 1.3 x 1 mm, rich brown or yellowish-brown, lustrous.

Distribution: Originally native to tropical Asia, now pantropically cultivated for the ornamental inflorescence, also naturalized; 11 collections of naturalized plants studied, all from the Guianas (GU: 5; SU: 4; FG: 2).

Selected specimens: Guyana: Pomeroon Distr., Moruka R., Mora Landing, de la Cruz 964 (NY, US); Upper Mazaruni R., de la Cruz 2126 (MO, NY, US); Northwest Distr., Waini R., de la Cruz 3751 (K, MO, NY, US). Suriname: Corantijnpolder near Nieuw-Nickerie, Lanjouw 633 (U). French Guiana: 10 km NW of Kourou, Roubik 114 (MO); Route Iracoubo-Mana, 16 km ouest d'Iracoubo, Descoings & Luu 20461 (CAY).

Vernacular names: Guyana: tocoroho. Suriname: staan vaste.

9. **IRESINE** P. Browne, Civ. Nat. Hist. Jamaica 358. 1756, nom. cons.
   Type: [specimen] P. Browne, Herb. Linnaeus No. 288.5 (typ. cons., LINN)

Erect or scrambling herbs to subshrubs; nodes tending to be flattened, encircled by a line of longer hairs. Leaves opposite, petiolate, often more or less pubescent. Inflorescence of axillary and terminal panicles of spikes; bracts and bracteoles present, persistent. Flowers minute, sessile, bisexual, or unisexual and dioecious; tepals 5, in female flowers with a tuft of short and straight, or long and tangled, hairs from base, in male flowers not hairy at base; stamens 5, filaments united below to form a cup, anthers 2-locular, pseudostaminodia present or not; ovule 1, stigmas 2(-3), subsessile.

Distribution: Approximately 80 species, indigenous to Australia and the New World tropics; 1 species in Guyana.

1. **Iresine diffusa** Humb. & Bonpl. ex Willd., Sp. Pl. 4: 765. 1806. Type: Peru, Humboldt s.n. (holotype B-W #18356; IDC 7440.1335: I.8,9, photo, not seen). – Fig. 23

   Iresine polymorpha Mart., Nov. Gen. Sp. Pl. 2: 56, t. 153. 1826. Type: not designated.

Erect to prostrate herb or scrambling vine to 2 m; stems glabrous to sparsely pubescent, frequently with a denser line of hairs at flattened nodes. Petiole 5-45 mm long; blade ovate, lanceolate or elliptical, 3-11 cm long, apex acute to acuminate, base rounded to cuneate, membranous, serrulate with minute ciliolulate teeth, sparsely pubescent above and beneath especially on veins. Inflorescence a diffuse panicle to ca. 45 cm long, comprised of numerous spikes having a "filigree" overall aspect; rachis of spikes very short, 2-3 mm long on ultimate, pubescent in lines; bracts ovate, 0.3-0.6 mm long, hyaline, shining, not keeled; bracteoles 0.5-0.8 mm long, hyaline, not keeled. Flowers unisexual, dioecious (rarely monoecious). Male flowers: tepals ovate to oblong, 0.8-1.3 mm long, stramineous- or white-hyaline, not clearly veined; stamens 5, filaments unequal. Female flowers: tepals oblong, 0.9-1.1 mm long, strongly 3-veined, stramineous- or white-hyaline, enveloped in wavy, white trichomes (lanate) about twice as long as tepals; ovule 1, stigmas 2(-3), linear, curved inwards. Seed lenticular, 0.5-0.6 mm long, reddish-brown, lustrous.

Distribution: Southern United States, tropical America; 6 collections studied, all from Guyana (GU: 6).

Selected specimens: Northwest Distr., Waini R., de la Cruz 1245 (MO, NY, US); Kamakusa, Upper Mazaruni R., de la Cruz 2774 (MO, US); Northwest Distr., Wanama R., de la Cruz 3970 (MO, NY, US).

10. **PFAFFIA** Mart., Beitr. Kenntn. Amarantaceen 103. 1825.
    Lectotype: P. glabrata Mart.

Perennial herbs, clambering vines or erect subshrubs, often with thick rootstock, usually pubescent; stem branched, often with interpetiolar ridges and a tuft of hairs in leaf-axils. Leaves opposite, simple, entire, pinnately veined, subsessile to petiolate, exstipulate. Inflorescence of dense terminal heads, or flowers distant and borne singly in short spikes

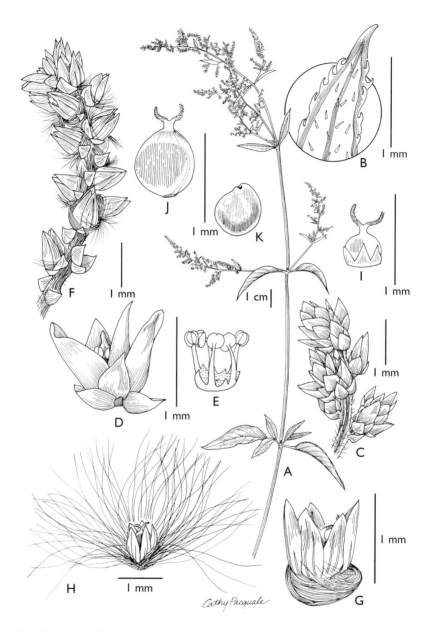

Fig. 23. *Iresine diffusa* Humb. & Bonpl. ex Willd.: A, habit; B, trichomes at leaf apex; C, male inflorescence; D, male flower; E, stamens; F, female inflorescence; G, female flower; H, female flower; I, pistil; J, fruit; K, seed. (A-B, de la Cruz 2774; C-E, de la Cruz 1245; F-K, de la Cruz 3785). Drawing by Cathy Pasquale.

arranged in diffuse, axillary or terminal, long-pedunculate cymes or panicles. Flowers bisexual, sessile, subtended by 1 ventral bract and 2 lateral bracteoles. Tepals 5, free, scarious, concave, subequal, outer 3 somewhat larger and distinctly longitudinally 3-ribbed (-veined), greenish-white, sometimes inside with a tuft of straight white hairs, exceeding perianth; stamens 5, filaments expanded and connate below in a cup, cup and filaments with or without apical appendages, anthers 2-locular, becoming 1-locular at anthesis, introrse, medifixed, pseudostaminodia absent; ovary ovoid or obovoid, ovule 1, campylotropous, placentation basal, with flattened, elongated funicle, stigma sessile, capitate or 2-lobed. Fruit a membranous, indehiscent capsule (sometimes designated a utricle) included in tepals; seed 1, lenticular, cochleate-orbicular, smooth.

Distribution: Approximately 30 species, occurring in tropical Central and South America from Mexico to Argentina; 3 species in the Guianas.

Literature: Stützer, O. 1935. Die Gattung Pfaffia. Repert. Spec. Nov. Regni Veg. Beih. 88: 1-46.

KEY TO THE SPECIES

1 Vining shrub to 10 m; flowers distant, borne singly at intervals along rachis of spikes; tepals shortly pubescent on the back, inside a large, dense tuft of long hairs that exceeds the perianth . . . . *3. P. grandiflora*
  Herb or shrub to 2 m; flowers densely aggregated at apex of a terminal, head-like, eventually elongated spike; tepals glabrous on the back, inside with a very small tuft of short hairs that is much shorter than the perianth 2

2 Leaves green; tepals 4.2-5.0 mm long; apical appendage of filaments well-developed . . . . *1. P. glabrata*
  Leaves bluish-green (glaucous); tepals 2-3 mm long; apical appendage of filaments not well-developed . . . . *2. P. glomerata*

1. **Pfaffia glabrata** Mart., Nov. Gen. Sp. Pl. 2: 21. 1826. Type: Brazil, Gardner 3965 (G-DC, not seen) (according to Tropicos).

In the Guianas only: var. **rostrata** O. Stützer, Repert. Spec. Nov. Regni Veg. Beih. 88: 24. 1935. Type: not designated.

Herb or subshrub to 2 m; stems ascending or erect, branched, 30-120 cm, glabrescent. Leaves petiolate; blade lanceolate or linear-lanceolate, 3.5-12 x 0.3-1.3(-2) cm, attenuate or acuminate at base and apex, glabrate above, strigose beneath. Inflorescence of simple, axillary or terminal, solitary head-like spikes; head-like spikes erect, ovoid-globose or subglobose, 3-5 cm long, eventually (becoming) elongated; bracts ovate-cordate, glabrous, slightly pilose. Tepals oblong-lanceolate, ca. 4.2-5 mm long, subacute, 3-veined, glabrous on back, with a very short tuft of hairs shorter than bracteoles at base; filaments ciliolate on margin, apical appendages of filaments oblong; stigmas subglobose.

Distribution: Peru, Brazil and French Guiana; 8 non-Guianan collections examined; no French Guiana specimens encountered. Stützer (1935) attributed 2 specimens from French Guiana to this variety: sin. loc., coll. 1826, Poiteau s.n. (K, P); near Cayenne, Poiteau s.n. (V).

Note: Distinctive for the large, 4.2-5 mm long tepals being glabrous on the back, and with a very short tuft of hairs from inside, not exceeding the perianth. Typical *P. glabrata* has leaves glabrate above and pubescent beneath, and tepals 4.2-5 mm long, whereas the characteristics given for French Guianan *P. glabrata* var. *rostrata* by Stützer are leaves pubescent on both sides and tepals 3 mm long, shorter than those (4.2-5 mm) examined by the present authors and which approach the characters of *P. glomerata*, although the latter has tepals more often 2 mm, rather than 3 mm long.

2. **Pfaffia glomerata** (Spreng.) Pedersen, Darwiniana 14: 450. 1967. – *Iresine glomerata* Spreng., Neue Entdeck. 2: 110. 1821. – *Serturnera glauca* Mart., Nov. Gen. Sp. Pl. 2: 37. 1826, nom. illeg. – *Pfaffia glauca* Spreng., Syst. Veg. 4(2): 107. 1827, nom. illeg. Type: Brazil (lectotype M) (designated by Pedersen, Darwiniana 1967: 451).

– Fig. 24

In the Guianas only: var. **glomerata**.

Gomphrena stenophylla Spreng., Syst. Veg. 1: 823. 1824. – *Pfaffia stenophylla* (Spreng.) Stuchlik, Repert. Spec. Nov. Regni Veg. 12: 357. 1913. Type: Uruguay: Montevideo, Sellow s.n. (holotype B) (according to Pedersen 1967: 451).
Pfaffia stenophylla (Spreng.) Stuchlik var. *foliosa* O. Stützer, Repert. Spec. Nov. Regni Veg. Beih. 88: 40. 1935. Type: not designated.

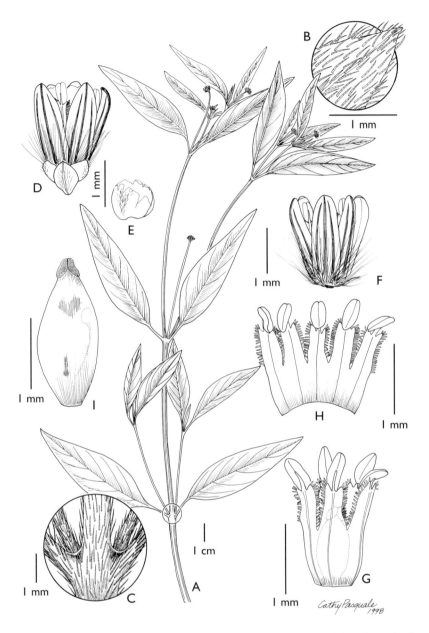

Fig. 24. *Pfaffia glomerata* (Spreng.) Pedersen: A, habit; B, trichomes at leaf apex; C, node; D, flower with basal bract and bracteoles; E, lateral bracteoles; F, tepals with basal hairs; G, androecium; H, stamens with opened cup; I, fruit. (A-H, Wittingthon 101; I, Oldeman B-3894). Drawing by Cathy Pasquale.

Herb or subshrub to 2 m; stem erect, dichotomously branched, glabrous below, pubescent above, subangular, striate, nodes more densely pubescent than internodes. Leaves bluish-green (glaucous), narrowly lanceolate, gradually becoming linear in upper part of stem, to 10 x 3.2 cm, acuminate at apex, somewhat cuneate at base, appressed-pilose above and beneath. Inflorescence a cyme of peduncled, subglobose heads of 4-8 mm wide, heads elongating to become floriferous for 1 cm; rachis pubescent; bracts broadly ovate, acute, 1/3 as long as tepals. Tepals oblong, 2-3 x ca. 0.5 mm, acute, 3-veined, glabrous; filaments with expanded portion scarcely lobed; ovary ovoid, stigma capitate. Seed reddish-brown, shining.

Distribution: Mexico and tropical South America; sometimes in disturbed areas of human habitation; 52 collections studied, 19 from the Guianas (GU: 2; SU: 2; FG: 15).

Selected specimens: Guyana: Ro. Schomburgk ser. I, 46 (NY, US); Potaro-Siparuni Region, Iwokrama Rainforest Reserve, Essequibo R., between Pisham Falls and Tiger Cr., Clarke 382 (US) . Suriname: Litanie, Stahel 88 (NY); Fleuve Maroni, Saint Goodou Campou, Sastre & Bell 8211 (U). French Guiana: Saül, Moretti 105 (CAY); Ile de Cayenne, Burgot 1 (CAY); Village Boni de Assici, Bassin du Maroni, Fleury 720 (CAY, U).

Uses: In French Guiana the plant is used to remedy fatigue (Burgot 1), and an infusion is employed to calm the nerves (Oldeman B-3894).

Vernacular names: Suriname: kaloeba kele (Arawak), oubelt (Carib). French Guiana: l'arbre sensible (Creole); booko kini, kini kini (Boni).

Note: Plants are said to be polygamo-dioecious (Smith & Downs, 1972).

3. **Pfaffia grandiflora** (Hook.) R.E. Fr., Ark. Bot. 16(12): 10. 1920. – *Iresine grandiflora* Hook., Icon. Pl. 2: ad t. 102. 1837. Type: not designated.

In the Guianas only: var. **grandiflora**.

Pfaffia grandiflora (Hook.) R.E. Fr. var. *typica* O. Stützer, Repert. Spec. Nov. Regni Veg. Beih. 88: 9. 1935, nom. inval.
Pfaffia grandiflora (Hook.) R.E. Fr. var. *hookeriana* (Hemsl.) O. Stützer f. *guianensis* Klotzsch ex O. Stützer, Repert. Spec. Nov. Regni Veg. Beih. 88: 10. 1935. Syntypes: Guyana: Barama R., Ri. Schomburgk 1527 (K).

Suffrutescent, clambering and vining perennial to 10 m, usually pubescent. Leaves lanceolate to broadly ovate, 2.0-13.5 x (1-)2.5-5.0 cm, acuminate at apex, rounded to acute at base, glabrous above, rufous-strigose beneath. Inflorescence a diffuse terminal panicle, overall to ca. 50 cm long, of pedunculate, to 8.1 cm long spikes; rachis and peduncles densely rufous-pubescent; flowers distant, borne singly at intervals along rachis of spikes; bracts and bracteoles subequal, broadly ovate, 1-2 mm long, rufous-tomentose or lanate. Tepals 5, subequal, elliptical, 3.0-4.3 x 2 mm, white, strigose on back, lanate inside from below with a tuft of straight hairs of 5 mm long and exceeding perianth, giving the open, expanded flower the appearance of being floccose within; stamens 5, basal tube exappendiculate; stigma capitate. Fruit included, to 2 mm long; seed reddish-brown, ca. 1.5 mm long.

Distribution: Puerto Rico, Mexico to South America; disturbed habitats; 26 collections studied, none from Guyana. Stützer (1935) cited the following specimens from Guyana: Barima R., Jenman 7120 (K); Barama R., coll. Oct. 1841-1844, Ri. Schomburgk 1527 (K).

34. PORTULACACEAE

by

ROBERT A. DEFILIPPS AND SHIRLEY L. MAINA[8]

Annual or perennial, succulent herbs or subshrubs; nodes glabrous or pilose. Leaves alternate, opposite or in a basal rosette, simple, often succulent, sessile or petiolate; stipules scarious, lacerate or fimbriate, or modified into hairs. Inflorescence a terminal or lateral raceme, cyme or panicle, or the flowers solitary and axillary. Flowers bisexual, regular; sepals 2(4-8), free or united below, imbricate, persistent or caducous, scarious or herbaceous; petals (2-)5(-12), free or united below, imbricate, deciduous; stamens 4-5 or more, free or adnate at base to corolla, filaments filiform, anthers 2-locular, basifixed, longitudinally dehiscent, introrse; ovary superior, half-inferior or inferior, 2- to 5-carpellate and 1-locular, placentation free-central or basal, ovules several to numerous, campylotropous, often on long funicles, styles and stigmas 1-9, free or united below. Fruit a loculicidal or circumscissile capsule (rarely an indehiscent nut); seeds (1-2 by abortion) 3-numerous, compressed, lenticular, reniform or cochleate, sometimes strophiolate, often tuberculate, embryo curved, perisperm often present and abundant, endosperm rudimentary, mealy.

D i s t r i b u t i o n : Approximately 450 species in 29 genera, essentially of worldwide distribution but mostly in the Northern Hemisphere; in the Guianas 7 species in 2 genera.

LITERATURE

Carolin, R.C. 1993. Portulacaceae. In K. Kubitzki, The Families and Genera of Vascular Plants 2: 544-555.

DeFilipps, R.A. 1992. Ornamental Garden Plants of the Guianas. An Historical Perspective of Selected Garden Plants from Guyana, Surinam and French Guiana. Portulacaceae. p. 175. Smithsonian Institution. Washington, D.C.

Eliasson, U.H. 1996. Portulacaceae. In G. Harling & L. Andersson, Flora of Ecuador 55: 29-53.

Geesink, R. 1969. An account of the genus Portulaca in Indo-Australia and the Pacific (Portulacaceae). Blumea 17: 275-301.

[8] National Museum of Natural History, Department of Systematic Biology - Botany, NHB 166, Smithsonian Institution, Washington, D.C. 20013-7012, U.S.A.

Grenand, P. *et al.* 1987. Pharmacopées traditionnelles en Guyane, Portulacaceae pp. 367-368.

Howard, R.A. 1988. Flora of the Lesser Antilles, Portulacaceae. 4: 200-207.

Legrand, C.M.D.E. 1952. Revisando tipos de Portulaca. Comun. Bot. Mus. Hist. Nat. Montevideo 2(24): 1-10 + 2 plates.

Legrand, C.M.D.E. 1962. Las especies americanas de Portulaca. Anales Mus. Hist. Nat. Montevideo, ser. 2. 7(3): 1-147 + 29 plates.

Lemée, A.M.V. 1955-1956. Flore de la Guyane Française, Portulacacées 1: 582-584. 1955; 4(1): 33. 1956.

Nevling, L.I. 1961. Portulacaceae. In R.E. Woodson & R.W. Schery, Flora of Panama 4: 427-431= Ann. Missouri Bot. Gard. 48: 85-89.

Ooststroom, S.J. van. 1943. Portulacaceae. In A.A. Pulle, Flora of Suriname 1(1): 486-491.

Ostendorf, F.W. 1962. Nuttige Planten en Sierplanten in Suriname, Portulacaceae. Bull. Landbouwproefstat. Suriname 79: 29-30.

KEY TO THE GENERA

1 Leaves usually with stipules fimbriate or modified into tufts of hairs; flowers solitary, or clustered at the apex of stems and branches, sessile or subsessile; ovary inferior or half-inferior, styles 1-12, free; capsule circumscissile (dehiscing by a lid transversely) . . . . *1. Portulaca*
Leaves exstipulate, without tufts of hairs; flowers in racemes or panicles, pedicellate; ovary superior; styles 3, united; capsule loculicidally 3-valved
. . . . *2. Talinum*

1. **PORTULACA** L., Sp. Pl. 445. 1753.
Lectotype: P. oleracea L.

Annual or perennial, prostrate to ascending, succulent herbs. Leaves alternate or subopposite, often upper ones crowded in a foliar involucre around flowers, flat or terete, sometimes succulent; stipules scarious, fimbriate or present as tufts of hairs, rarely absent. Inflorescences of subapical flowers, clustered in a capituliform cyme or solitary. Sepals 2, opposite, free or connate below, unequal, usually persistent; petals (4-)5(-6), free or connate at base, falling off rapidly; stamens 4-8 or more, basipetalous, filaments often pubescent below; ovary inferior of half-inferior, ovules numerous, styles 2-9, united below, rarely free. Fruit a 1-locular, membranous or chartaceous, circumscissile capsule; seeds numerous, reniform or cochleate, testa smooth or tuberculate.

D i s t r i b u t i o n : Approximately 200 species, distributed in temperate to tropical regions worldwide, often in dry areas; 5 species in the Guianas, sometimes planted as ornamentals.

KEY TO THE SPECIES

1 Stems weak, filiform (0.5 mm wide), creeping, rooting at the nodes; leaves opposite; petals yellow . . . . *4. P. quadrifida*
Stems stout, often fleshy, erect or spreading, not rooting at the nodes; leaves alternate; petals white, yellow, pink or purple 2

2 Leaves obovate or spathulate, flat; nodes glabrous or with few stipular hairs; petals yellow . . . . *2. P. oleracea*
Leaves linear to linear-oblong, terete or slightly flattened; nodal hair clusters usually conspicuous; petals white, pink or purple 3

3 Flowers often double; petals 1.5-2.5 cm long; seeds grey . . . .
. . . . *1. P. grandiflora*
Flowers simple; petals 3-6 mm long; seeds grey or black 4

4 Leaves 5-16(-27) x 1-4 mm; petals rose-purple; seeds black . . . *3. P. pilosa*
Leaves 3-6 x 1-1.5 mm; petals white or whitish-rose; seeds grey to black . .
. . . . *5. P. sedifolia*

1. **Portulaca grandiflora** Hook., Bot. Mag. 56: ad t. 2885. 1829. –
Portulaca pilosa L. subsp. *grandiflora* (Hook.) R. Geesink, Blumea
17: 297. 1969. Type: [icon] Hook., Bot. Mag. 56: t. 2885. 1829
(according to Howard 1988: 202).

Annual herb; stems ascending or prostrate, to 30 cm long, with nodal tufts of hairs. Leaves alternate; blade fleshy, terete or linear-cylindrical, 5-30 x 1-3 mm, obtuse, acute or acuminate at apex. Inflorescence a terminal cluster of 1-3 flowers, surrounded by long (2-3 mm) white or brownish hairs, and an involucre of 6-9(-12) leaves. Flowers to ca. 2.5 cm wide, often double (in horticulture); sepals unequal, deltoid-ovate, 6-10 x 6-8 mm, acute at apex; petals obovate, 1.5-2.5 x 1.5-2.5 cm, sometimes apically notched, whitish (wild), pink, salmon, purple, red or yellow, sometimes striped; stamens numerous, filaments filiform, to 7 mm long, anthers red, ca. 0.8 mm long; style to 1 cm long, stigmas ca. 10 through rebranching of style. Fruit conical, broadly ellipsoid or subglobose, 4-5 x 3-4.5 mm, circumscissile slightly below middle; seeds grey, tuberculate, metallic-iridescent.

Distribution: Argentina, Uruguay; large- and double-flowered forms are grown as ornamentals at the Botanic Gardens, Georgetown, Guyana, at the Esther Stichting near Paramaribo, and elsewhere (Ostendorf, 1962) in Suriname, and at the Jardin Botanique, Cayenne, French Guiana (DeFilipps, 1992); 15 collections studied, 3 from the Guianas (GU: 1; FG: 2).

Selected specimens: Guyana: Rupununi Distr., Dadanawa, Jansen-Jacobs *et al.* 3886 (US). French Guiana: Pointe de Bourda, Ile de Cayenne, Hoff 5106 (CAY, US); Jardin creole, Cayenne, Prévost 1325 (CAY).

Vernacular names: Suriname: tienuursklokje, portulak (Dutch). French Guiana: chevalier de onze heures.

Note: *Portulaca grandiflora* Hook. occurs as a native plant in Argentina and Uruguay, and it also has been horticulturally improved into large-petaled, double-flowered cultivars with various flower colors. French Guianan herbarium specimens seem to resemble or represent two entities: 1) the var. *immersostellulata* (Poelln.) D. Legrand, which is grown in gardens [e.g., Prévost 1325 from French Guiana]; and 2) the wild var. *grandiflora* f. *depressa* (D. Legrand) D. Legrand, which differs from *immersostellulata* by seed-coat ornamentation characters, and is cited from Argentina by Legrand (1962); it resembles a robust *P. pilosa* with linear leaves.

2. **Portulaca oleracea** L., Sp. Pl. 445. 1753. Type: Loefling, Herb. Linnaeus No. 625.1 (lectotype LINN) (designated by Geesink 1969: 292). – Fig. 25

Portulaca oleracea L. var. *granulatostellata* Poelln., Occas. Pap. Bernice Pauahi Bishop Mus. 12(9): 5. 1936. Type: Hawaii, Oahu, Hosaka 419 (lectotype BISH, not seen) (designated by Geesink 1969: 293).

Annual herb, succulent; fibrous-rooted; stems prostrate or decumbent, spreading to 30(-50) cm long, purplish, glabrous or with few nodal hairs of ca. 1 mm long. Leaves alternate, upper leaves often opposite; stipules inconspicuous, minutely fimbriate or absent; petiole 1-8 mm long; blade fleshy, flat, elliptical, obovate or spathulate, 0.6-4 x 0.2-2 cm, obtuse or truncate at apex, cuneate at base, glabrous. Inflorescence of terminal, solitary or clustered, sessile or subsessile flowers, clusters of up to 10 fowers. Sepals green, broadly ovate to orbicular-ovate or triangular, 2.8-5 x 2.8-4 mm, broadly keeled above, united at base, usually persistent; petals yellow, 4-5, obovate, 3-8 x 1.5-3 mm, ephemeral; stamens 6-20, filaments 1.5-1.75 mm long, anthers broadly oblongoid or globose, 0.4-0.5 mm long; ovary half-inferior, ovoid or short-conical, stigmatic branches 4-6. Fruit broadly ovoid to fusiform, ca. 4-5 x ca. 2.5 mm, circumscissile from ca. 1/3 up from base, to near middle; seeds black, cochleate, ca. 0.5-0.8 mm wide, very finely and minutely granulate or tuberculate.

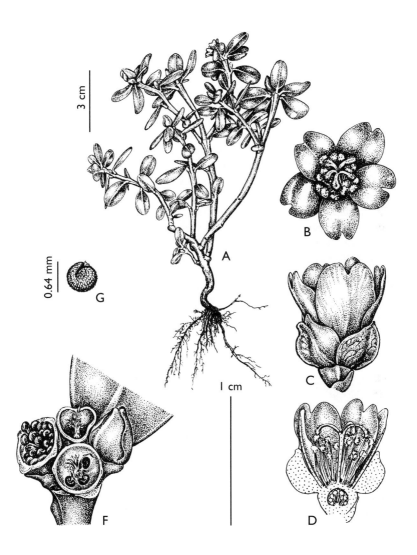

Fig. 25. *Portulaca oleracea* L.: A, habit, x 2/3; B, flower, from above, x 4; C, flower, side view, x 4; D, flower longitudinally dissected, x 4; F, node with dehisced fruits, x 4; G, seed, x 14. Drawing by P. Fawcett, reprinted from Correll, D.S. & H.B. Correll. 1982. Flora of the Bahama Archipelago.

Distribution: Subcosmopolitan weed, indigenous to the Old World tropics and introduced elsewhere; in the Guianas on disturbed ground; 79 collections studied, all from the Guianas (GU: 20; SU: 39; FG: 20).

Selected specimens: Guyana: Kanuku Mts., Nappi-head on Nappi Cr., Camp 1, Jansen-Jacobs *et al.* 665 (CAY, U); Rupununi R., Monkey Pond landing, SW of Mt. Makarapan, Maas *et al.* 7635 (B, U); Pomeroon Distr., Waramuri Mission, Moruka R., de la Cruz 2564 (NY). Suriname: Paramaribo, way to Kwatta, Samuels 330 (K); Along road from Nieuw Nickerie to the sea, Wessels Boer 529 (U); Lower Saramacca R. near Plantation Catharina Sophia, Lanjouw 277 (U). French Guiana: Maripasoula, Sauvain 623 (CAY); Route de Baduel, Ile de Cayenne, Hoff 5709 (CAY); Iles de Salut, Ile Royale, Cremers 8433 (CAY).

Uses: Sometimes cultivated as a leafy vegetable in Suriname, and the leaves also mixed with sugar or soap and used for ripening abscesses (Ostendorf, 1962). In French Guiana, the Palikur Amerindians crush the leaves and stems in water, and drink the resulting liquid as an hypotensive, whereas French Guianan Creoles use a tea of the plant as an antidiabetic (e.g., Ducatillon & Gelly 49) and digestive. They also use the whole plant as an emollient for muscular aches and make a purgative and a drink for albuminuria from the plant (Grenand *et al.*, 1987).

Vernacular names: Guyana: hog bhajee. Suriname: piendjalang (Carib); postelein (Dutch); porselein, gron-possie, gron-posren (Sranan); lonia (Hindi); krokot (Javan); bembe, pose (Saramaccan). French Guiana: pourpier; croupier, croupier blanc (Creole); akusinami (Wayapi); posin (Boni); posen (Ndjuka).

Note: No seemingly authentic Guianan specimens of *Portulaca mucronata* Link, (cited for the Guianas, e.g. by Lemée (1955) and Grenand (1987) for French Guiana), have been seen in this study; it is a South American yellow-flowered plant with mucronate leaves, resembling *P. oleracea*.

3. **Portulaca pilosa** L., Sp. Pl. 445. 1753. Type: Herb. Linnaeus No. 625.2 (LINN) (according to Howard 1988: 203).

Portulaca lanata Rich., Actes Soc. Hist. Nat. Paris 1: 109. 1792. Type: not designated.
Portulaca pilosa L. var. *guyannensis* D. Legrand, Anales Mus. Hist. Nat. Montevideo ser. 2. 7(3): 83. 1962. Type: Brazil, Amazonas, Spruce 2256 (holotype NY, not seen; isotypes GH, CAS, not seen).

Annual herb, succulent; stems ascending or prostrate, to 20-30 cm long, glabrous, but usually with conspicuous nodal tufts of 3-7 mm long, white hairs. Leaves alternate, sessile or subsessile; blade terete, somewhat flattened, linear, linear- or oblong-lanceolate, 5-16(-27) x 1-4 mm, glabrous. Inflorescence of terminal, solitary or clustered, sessile flowers, surrounded by long brownish or whitish hairs and an involucre of 6-10 leaves. Sepals somewhat unequal in size and shape, triangular-ovate to ovate, 2-5 mm long, not keeled, acute, apiculate; petals rose-purple, reddish or purple-pink, obovate, oblong-obovate, or broadly ovate, 3-6 x 2.5-4.5(-6) mm, sometimes retuse; stamens 15-37, filaments crimson, 2-2.5 mm long; style 2-2.5 mm long, stigmas 3-6. Fruit ovoid or subglobose, to 7 x 2.5-4.3 mm, rich yellowish-brown and glossy above, circumscissile at about middle; seeds lenticular-reniform, 0.5-0.7 mm wide, black, dull or shining, minutely tuberculate.

Distribution: Southern United States and New World tropics; often as a weed in open places; 21 specimens examined, all from Guyana and French Guiana (GU: 9; FG: 12).

Selected specimens: Guyana: Kanuku Mts., Rupununi R., Bush Mouth near Witaru Falls, Jansen-Jacobs *et al.* 128 (CAY); Pomeroon Distr., Moruka R., Santa Rosa, de la Cruz 996 (NY); Takutu R., Graham 371 (K). French Guiana: Kourou, savane Matiti, Sastre *et al.* 4208 (CAY); Ile de Cayenne, rochers de la pointe de Bourda, de Granville 7261 (U); Iles du Salut, Ile Royale, Cremers 8451 (CAY).

Use: Sometimes cultivated in Suriname gardens (Ostendorf, 1962).

Notes: The characteristic "pilose" appearance of the stems is due to the very long, profusely fimbriate stipules at the nodes, which often curl and overlap to a profound degree.
K. von Poellnitz (Repert. Spec. Nov. Regni Veg. 37: 240-320. 1934), with some uncertainty, synonymized the yellow-flowered *P. rubricaulis* Kunth into *P. pilosa* L., whereas *P. rubricaulis* was accepted as a good species by Legrand (1962), who cited a Sagot specimen from French Guiana; the specimen has not been seen by the present authors, who would defer acceptance of the species for the Guianas until a modern revision of the genus is available.
Similarly questionable is a French Guiana *Portulaca* specimen (Broadway 962, US) determined by Legrand as *P. teretifolia* Kunth and having tuberculate seeds (rather than non-tuberculate as stated in his own 1962 species description), while other specimens [Broadway 105 (US) and Broadway 962 (US)] supposedly *P. teretifolia,* have stems that closely resemble those of *P. pilosa.*

4. **Portulaca quadrifida** L., Mant. Pl. 73. 1767. Type: not designated.

Annual herb; stems prostrate, much-branched, to ca. 14 cm long, weak and filiform, 0.5 mm wide, glabrous, creeping and rooting at nodes, forming rooting mats, with nodal tufts of to 2 mm long hairs. Leaves opposite; blade flat, obovate, elliptical, lanceolate or ovate, 2-12 x 1-6 mm (typically 4 x 2 mm in the Guianas), obtuse or acute at apex. Inflorescence of 1-2(-3) terminal flowers, surrounded by long white hairs and an involucre of 4-5 triangular, obtuse, membranaceous, 2-3 mm long leaves; flowers pedicellate. Sepals subequal, triangular-oblong, 2-3.5 x 2 mm, apex obtuse-rounded; petals yellow, elliptical or ovate, 3-6 mm long; stamens 7-12, filaments pilose, anthers more or less suborbicular; ovary conical-ovoid, style 2.5-3.5 mm long, 3- to 4-branched, stigmas ca. 0.5 mm long. Fruit ovoid, to 6 x 0.5 mm, circumscissile in lower third, operculum tubular-campanulate, 1.5-2.5 mm wide; seeds bluish-grey, 0.8-1.0 mm wide, densely granular- or spiny-tuberculate.

Distribution: Cosmopolitan, in the New World in the West Indies and South America; 4 specimens examined, all from Guyana (GU: 4).

Selected specimens: Guyana: Georgetown, Promenade Gardens, Hitchcock 16606 (US); Georgetown, Jenman 4991 (K); Coast lands, Jenman 5311 (K photo, U).

5. **Portulaca sedifolia** N.E. Br., Trans. Linn. Soc. London, Bot. ser. 2. 6: 19. 1901. Type: Brazil-Guyana frontier, Ireng R., McConnell & Quelch 237 (not seen).

Portulaca cayennensis D. Legrand, Comun. Bot. Mus. Hist. Nat. Montevideo 2(22): 2. 1952. Type: French Guiana, around Cayenne, Broadway 16 (NY, not seen).

Annual subshrub or herb; stems prostrate or ascending, branched from base, 4-10 cm long, purplish or reddish, nodes with tufts of long, whitish hairs as long as or longer than the leaves. Leaves alternate, succulent; shortly petiolate; blade flat, elliptic-oblong, linear-oblong or linear-lanceolate, 3-6 x 1-1.5 mm, acute or obtuse, glabrous, purplish or reddish, often papillose or with a corrugated surface. Inflorescence a terminal, solitary flower or few-flowered cluster subtended by 5-6 broadly ovate, ca. 1.5 mm long, concave involucral leaves. Sepals deltate or ovate-triangular, 1.5-3 mm long, obtuse, not keeled; petals white or whitish-rose, sometimes yellowish-white or pink; stamens 5-15; style 1 mm long, style-branches 2-5. Fruit ovoid, 2-3.5 mm, circumscissile below middle, operculum brownish-yellow, shining; seeds cochleate-orbicular, 0.5-0.65 mm, black, slate-grey or greyish-black, tuberculate.

Distribution: Northern Brazil, Venezuela, and the Guianas; in saxicolous, dry, exposed habitats, often at the seashore or on granitic rocks; 21 collections studied, all from the Guianas (GU: 5; SU: 1; FG: 15).

Selected specimens: Guyana: Rupununi Distr., Shea Rock, Jansen-Jacobs *et al*. 3671, 4806 (US); Rupununi Distr., Dadanawa, Tawatawun Mt., Jansen-Jacobs *et al*. 3979 (US); U. Takutu-U. Essequibo Region, Rupununi Savanna, Towatawan Mt., 11 km E of Dadanawa, Gillespie 1976 (US). Suriname: Islands in rapids between Grasi Falls and Posoegronoe, Maguire 23996 (US). French Guiana: Region de la Haute Crique Armontabo, bas Oyapock, de Granville 4339 (CAY); Haut Tampoc, Saut Pier Kourou, Cremers 4642 (CAY); Pripris Maillard, 4 km SE de Tonate, Raynal-Roques 20042 (CAY).

Note: Individual plants sometimes have dimorphic leaves, with some of the upper branches bearing flat leaves and some of the lower branches having prominently ellipsoid-, ovoid-, or subgloboid-inflated, 2-2.5 mm long, leaves.

2. **TALINUM** Adans., Fam. Pl. 2: 245. 1763, nom. cons.
   Type: T. triangulare (Jacq.) Willd. (Portulaca triangularis Jacq.), typ. cons. [= T. fruticosum (L.) Juss.]

Herbs, subshrubs or shrubs, often succulent and with fleshy tuberous roots. Leaves alternate (or subopposite), fleshy, entire, flat or terete, petiolate; stipules absent. Inflorescences a terminal or axillary raceme, panicle or cyme, or the flowers solitary and axillary. Sepals 2, free, caducous; petals 5, free or united at base, falling off rapidly; stamens 5-30, arranged in fascicles; ovary superior, sessile or short-stipitate, ovules numerous, basal, style 1, filiform, 3-branched at apex. Fruit a 1-locular, loculicidally 3-valved capsule, splitting from apex to base, chartaceous; seeds numerous, flattened, orbicular to reniform, shining, with evanescent aril, embryo incompletely annular; testa smooth, striate or tuberculate.

Distribution: Approximately 40 species in warm areas around the world; 2 species in the Guianas.

Literature: Poellnitz, K. von. 1934. Monographie der Gattung Talinum Adans. Repert. Spec. Nov. Regni Veg. 35: 1-34.

KEY TO THE SPECIES

1 Inflorescence a raceme or subcorymbose cyme to 19 cm long; pedicels mostly ascending; petals 6-14 mm long . . . . *1. T. fruticosum*
  Inflorescence a panicle to 60 cm long; pedicels spreading; petals 3-5 mm long . . . . *2. T. paniculatum*

1. **Talinum fruticosum** (L.) Juss., Gen. Pl. 312. 1789. – *Portulaca fruticosa* L., Syst. Nat. ed. 10. 1045. 1759. Type: [icon] Plum., Pl. Amer. 6: t. 150, f. 2. 1757 (designated by Wijnands & Westphal-Stevels, Taxon 34: 309. 1985).

Talinum triangulare (Jacq.) Willd., Sp. Pl., ed. 4. 2: 862. 1799. – *Portulaca triangulare* Jacq., Enum. Syst. Pl. 22. 1760. Type: [icon] Plum., Pl. Amer. 6: t. 150, f. 2. 1757 (designated by Wijnands & Westphal-Stevels, Taxon 34: 309. 1985) (homotypic synonym of *Portulaca fruticosa* L. by means of lectotypifications).

Annual or short-lived perennial herb; stems erect, stout, fleshy, glabrous, to 1 m tall, often from a woody root. Leaves alternate, often deciduous in the dry season; petiole to ca. 1 cm long; blade fleshy, oblanceolate to obovate, 2-14 x 0.6-4.5 cm, usually rounded or emarginate at apex, attenuate or cuneate at base, glabrous. Inflorescence a few- to many-flowered raceme or subcorymbose cyme, to 7(-19) cm long; pedicels 3-angled, 7-11 mm long, mostly ascending. Sepals pinkish-green, free, suborbicular, broadly ovate or lance-ovate, 5-7 mm long, cuspidate, persistent; petals white, pink or purple, obovate, broadly elliptical or ovate, 6-14 x 6 mm; stamens 20-40, filaments 4-5 mm long, anthers 0.8-1 mm long; ovary subglobose, ca. 1.5 mm wide. Fruit globose or subglobose, 4-6 mm wide; seeds flattened-orbicular, ca. 1 mm wide, black, shining, striolulate in concentric rows.

Distribution: Southern United States, New World tropics, including the Guianas; 2 collections seen from Guyana (GU: 2).

Selected specimens: Guyana: Upper Takutu-Upper Essequibo Region, Lethem, Clarke 2100 (US); East Berbice-Corentyne Region, Orealla, Corentyne R., Mutchnick 1182 (US).
Specimens from literature: van Ooststroom (1943) cites from Suriname: Versteeg 521; Pulle 223, 1449; Lanjouw 693; Stahel Wilh. Exp. 13, Lanjouw 1053. Specimens at U include Kramer & Hekking 2379, Everaarts 806, and Jonker 258 from Suriname; and Jenman 4000 from Guyana (all not seen).

Use: In Guyana and Suriname sometimes cultivated as a vegetable (Ostendorf, 1962).

Vernacular names: Suriname: Surinaamse postelein, postelein, bokolille, posren (Sranan).

2. **Talinum paniculatum** (Jacq.) Gaertn., Fruct. Sem. Pl. 2: 219. 1791.
   – *Portulaca paniculata* Jacq., Enum. Syst. Pl. 22. 1760. Type: not
   designated.   – Fig. 26

Herb to 2 m tall, sometimes woody at base, often tuberous-rooted, glabrous; stems green flushed with purple. Leaves alternate, succulent; petiole to ca. 1 cm long glabrous; blade obovate or broadly elliptical, 2.7-13 x 1.5-5.5 cm, obtuse or acute at apex, cuneate or tapering to base, glabrous. Inflorescence a terminal, cymosely much-branched panicle, 7-60 x 6-20 cm, bracteate; pedicels 6-25 mm long, somewhat dilated distally, spreading. Sepals pale green, ovate to orbicular, 2-4 x 1-2 mm, connate at base, slightly keeled, becoming reflexed, caducous; petals pale yellow, mauve or pink, elliptical, ovate or orbicular, 3-5 x 1-3 mm; stamens 10-ca. 20, filaments 1.5-2.0 mm long, violet, free, anthers ovoid, 0.5 mm long, yellow; ovary sessile, globose, ca. 1 mm long. Fruit ovoid to globose, 3-5 mm wide; seeds flattened-orbicular, 0.75-1.0 mm wide, rich dark brown to black, shiny, striolulate in concentric rows.

Distribution: Southern United States, New World tropics, including the Guianas; often on rocks near watercourses or the sea; 14 collections studied, all from the Guianas (GU: 2; SU: 1; FG: 11).

Selected specimens: Guyana: Potaro-Siparuni Region, Northern Pakaraimas, Kato Village, Koa R., Mutchnick 1573 (US); Kanuku Mts., Nappi-head on Nappi Cr., Camp 1, Jansen-Jacobs *et al.* 667 (CAY). Suriname: Wia-wia Reserve, Julen LBB 14557 (U). French Guiana: Pointe de Bourda, Ile de Cayenne, Bordenave 106 (CAY); Estuaire de la crique Malmanoury, Billiet & Jadin 4738 (CAY); Iles du Salut, Ile Royale, Cremers 8499 (CAY).

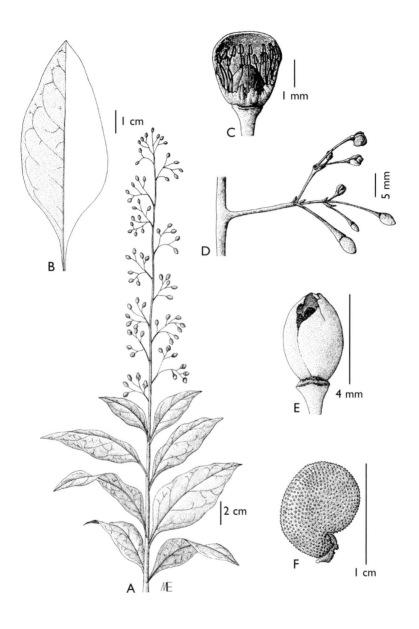

Fig. 26. *Talinum paniculatum* (Jacq.) Gaertn.: A, leaves and inflorescence; B, leaf venation; C, dissected flower; D, infructescence; E, fruit; F, seed. Drawing by M. Escamilla, reprinted from Ford, D.I. 1986. Flora de Veracruz 51.

35. BASELLACEAE

by

ROBERT A. DEFILIPPS AND SHIRLEY L. MAINA[9]

Perennial herbs, or shrubby trailing or climbing vines; stems slender, or stout and succulent, glabrous; rhizomes sometimes with fleshy tubers. Leaves alternate, sometimes succulent, mucilaginous, simple, entire; sessile or petiolate; stipules absent. Inflorescence of axillary or terminal spikes, racemes or panicles; peduncles nearly absent to present; pedicel subtended by a thin bract; flower always subtended by 2 or 4 deciduous or persistent, sometimes connate bracteoles. Flowers bisexual, actinomorphic, sessile or pedicellate; tepals (4-)5(-13), green, sometimes white to purple, entire, forming a calyx-like perianth; stamens 5, opposite tepals, anthers 4-locular, dehiscent by lateral slits; ovary superior or half-inferior, 3-carpellate, 1-locular, ovule solitary, basal, erect, campylotropous to orthoamphitropous, styles 3, stigmas 1 or 3. Fruit a dry utricle or nutlet, sometimes baccate and fleshy, enclosed by accrescent, sometimes winged perianth; seed 1, embryo annular, cochleate or spiral, perisperm sparse.

Distribution: Approximately 20 species in 4 genera in the New and Old World tropics, most diverse in South America; in the Guianas 2 species in 2 genera.

LITERATURE

Bogle, A.L. 1969. The genera of Portulacaceae and Basellaceae in the southeastern United States. J. Arnold Arbor. 50: 566-598.

Eriksson, R. 1996. Basellaceae. In G. Harling & L. Andersson, Flora of Ecuador 55: 55-83.

D'Arcy, W.G. 1979. Basellaceae. In R.E. Woodson & R.W. Schery, Flora of Panama 4. Ann. Missouri Bot. Gard. 66: 109-116.

Kellogg, E.A. 1988. Basellaceae. In R.A. Howard, Flora of the Lesser Antilles 4: 207-210.

Ostendorf, F.W. 1962. Nuttige Planten en Sierplanten in Suriname, Basellaceae. Bull. Landbouwproefstat. Suriname 79: 30.

Sidwell, K. 1999. Typification of two Linnaean names in the Basellaceae. Novon 9: 562-563.

Sperling, C.R. and V. Bittrich. 1993. Basellaceae. In K. Kubitzki, The Families and Genera of Vascular Plants 2: 143-146.

[9] National Museum of Natural History, Department of Systematic Biology - Botany, NHB 166, Smithsonian Institution, Washington, D.C. 20013-7012, U.S.A.

122

KEY TO THE GENERA

1 Flowers pedicellate, in simple or 1- to 3-branched slender racemes, opening
wide at anthesis; fruit a utricle; rhizomes sometimes tuber-bearing . . . . . .
. . . . *1. Anredera*
Flowers sessile, in fleshy spikes, usually cleistogamous or partly closed
around the anthers at anthesis; fruit baccate, juicy; rhizomes without
tubers . . . . *2. Basella*

1. **ANREDERA** Juss., Gen. Pl. 84. 1789.
Type: A. spicata J.F. Gmel. [= A. vesicaria (Lam.) C.F. Gaertn.]

Twining vines; rhizomes sometimes with fleshy tubers. Leaves petiolate;
blade ovate or elliptical, glabrous, sometimes succulent. Inflorescences
axillary (sometimes terminal), pendent, of sparingly (1- to 3-)branched
racemes, lax; pedicels short, slender; bract linear; flower subtended by 2
bracteoles. Tepals basally connate, white; stamens 5, inserted at base of
tepals, filaments white, hyaline, flattened, curved in bud, anthers oblong,
versatile or basifixed; styles and stigmas 3. Fruit a globose utricle
enclosed by perianth and accrescent bracteoles.

Distribution: Approximately 10-15 species in New World temperate
and tropical areas, introduced elsewhere; 1 species in French Guiana.

1. **Anredera leptostachys** (Moq.) Steenis, Fl. Males. Ser. 1. 5: 302.
1957. – *Boussingaultia leptostachys* Moq. in A. DC., Prodr. 13(2):
229. 1849. Syntypes: Mexico, Andrieux s.n. (not seen); Puerto Rico,
West s.n. (G-DC, not seen). – Fig. 27

Glabrous vine; stems climbing to 8 m. Petiole 0.5-1 cm long; blade
elliptical, lanceolate or ovate, to 7(-11) x 5 cm, acute to acuminate,
decurrent on petiole, cuneate at base. Racemes very slender, 1- to 3-
branched, to 40 x 0.8 cm, lax, axis ca. 1 mm wide; bracts linear-
acuminate or linear-subulate, 0.7-1.9 mm long, hyaline, caducous;
pedicels 0.7-1.5 mm long; bracteoles shorter than perianth, triangular-
subulate or ovate, 0.4-0.8 mm long, concave, hyaline. Flower strongly
scented; tepals white, connate for most of their length, oblong to ovate,
7.5 mm long, obtuse or rounded, spreading at anthesis; stigmas deeply
cleft. Utricle obovoid, ribbed, smooth, pale brown and shining below,
darker brown and rugose above.

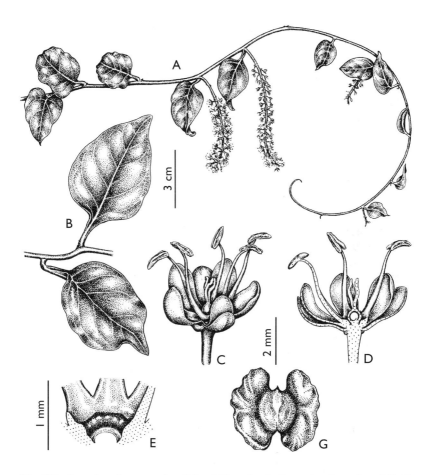

Fig. 27. *Anredera leptostachys* (Moq.) Steenis: A, flowering branch, x 2/3; B, section of leafy stem, x 2/3; C, flower, side view, x 8; D, flower longitudinally dissected, x 8; E, base of filaments, x 24; G, fruit, x 8. Drawing by P. Fawcett, reprinted from Correll, D.S. & H.B. Correll. 1982. Flora of the Bahama Archipelago.

Distribution: Florida, Texas, West Indies, Central and northern South America, including French Guiana; 2 collections studied (FG: 2).

Specimens examined: French Guiana: Ile de Cayenne, Commune de Remire, Wittingthon 89 (CAY); Cayenne, Jacquemin 2602 (CAY).

Vernacular name: French Guiana: glycerine (Creole).

2. **BASELLA** L., Sp. Pl. 272. 1753.
   Lectotype: B. rubra L. [= B. alba L.]

Glabrous vines or herbs; stems sometimes succulent; rhizomes without tubers. Leaves sessile or shortly petiolate, succulent. Inflorescences terminal or axillary, erect spikes or panicles; rachis often thick, fleshy; flowers sessile; bracteoles 4, unequal. Flowers almost always cleistogamous; tepals 5, pink, red or white, basally connate, imbricate, urceolate, fleshy and enlarging in fruit; stamens 5, filaments straight in bud, anthers dorsifixed, extrorse; styles 3. Fruit baccate, fleshy, maroon or black; seed subglobose.

Distribution: About 5 species in Africa and Asia, introduced elsewhere, including 1 species in the Guianas.

1. **Basella alba** L., Sp. Pl. 272. 1753. Type: Nepal, Mahakali Zone, Kanchapur Distr., W of Dhangarhi, Nicolson 2848 (neotype BM; isoneotype US) (designated by Sidwell 1999: 563). – Fig. 28

   Basella rubra L., Sp. Pl. 272. 1753. Type: Drawing of fruiting plant in Herb. Hermann 5, t. 207, No.119 (lectotype BM, not seen) (designated by Verdcourt in Milne-Redhead & Polhill, Fl. Trop. E. Africa, Basellaceae 2. 1968).

Glabrous herb, becoming a slender, twining vine; stems green to red, at first stout, to 2 cm wide, later becoming tall and climbing to ca. 10 m long. Petiole to 8 cm long, sometimes absent; blade broadly ovate to elliptical, to 15 x 15 cm, undulate, apex acute to obtuse or rounded, base cuneate to truncate or cordate. Inflorescence of axillary or subterminal spikes, to 26 cm long; bracts broadly ovate to lanceolate, 1.1-2.3 mm long, acuminate, hyaline, 1-veined; flowers sessile; bracteoles calyx-like, 1-2 mm long, acute. Flowers not scented; tepals red, pink or white, united to above middle, urceolate to cylindrical, oblong to ovate, 2.0-5.2 x 2-2.5 mm long, obtuse; anthers oblong, sometimes subsagittate at base. Baccate drupe globose, when fresh to 7 mm diam., enveloped by enlarged, succulent perianth, dark purple or shining black, with violet juice; seed globose, 4 mm diam.

Distribution: Originally native to the Old World (probably Africa and Southeast Asia), now of pantropical occurrence, including the Guianas; cultivated in gardens as a spinach-like potherb; 8 collections examined, all from the Guianas (GU: 5; SU: 2; FG: 1).

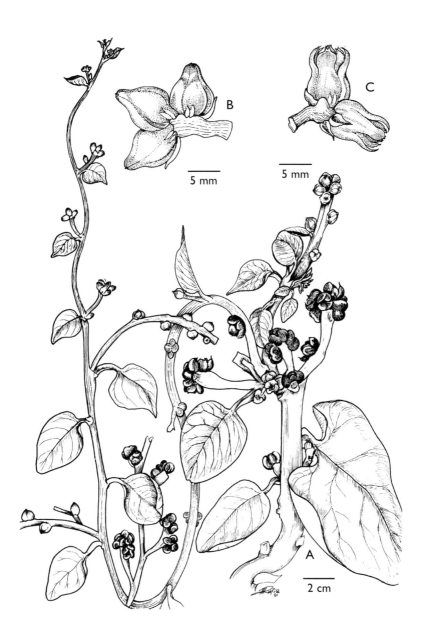

Fig. 28. *Basella alba* L.: A, flowering branch; B, cleistogamous flowers; C, open flowers. Reprinted from Malik, K.A. 1984. Basellaceae. In E. Nasir & S.I. Ali, Flora of Pakistan 161.

Selected specimens: Guyana: East Coast Demerara, Mon Repos, Omawale & Persaud 57 (NY); Potaro, 10 mi S of Potaro Landing, Hitchcock 17401 (NY). Suriname: Distr. Nickerie, Nieuw-Nickerie, Hekking 1132 (U); Paramaribo, Palmentuin, Kramer & Hekking 2861 (U). French Guiana: Cayenne, Prévost 1482 (CAY).

Use: Grown for the leaves, which are eaten as a spinach-like potherb in Guyana and Suriname (Ostendorf, 1962), and in French Guiana (Prévost 1482).

Vernacular names: Guyana: australian poi; green stem poi; purple stem poi; poisang; bambee. Suriname: spinazie (Dutch), poi (Hindi), gendola (Malay). French Guiana: epinard pays (Creole). Outside the Guianas the plant is known as "ceylon spinach" and "malabar spinach".

36. MOLLUGINACEAE
by
ROBERT A. DEFILIPPS AND SHIRLEY L. MAINA[10]

Annual or perennial, glabrous herbs or subshrubs, with monopodial or sympodial branching. Leaves alternate, opposite or whorled, simple, entire, often somewhat fleshy and crowded into a basal rosette and with pseudowhorls on stem; exstipulate or stipules membranous, sometimes laciniate; petiolate (sometimes weakly). Inflorescence mostly terminal or seemingly axillary cymes, sometimes few-flowered axillary fascicles or cymes, or solitary flowers; bracts and bracteoles minute. Flowers bisexual (rarely unisexual and dioecious), regular; sepals (4-)5, free, membranous; petals absent in Guianan species or few to numerous; stamens (3-)4-5(-numerous), filaments usually connate at base in short cupular tube, anthers 4-locular, dehiscing by longitudinal slits; ovary superior, (1-)2- to 5-locular, ovules 1-numerous per locule, placentation axile, anatropous to campylotropous, bitegmic, crassinucellate, styles or stigmas as many as locules, usually free. Fruit a loculicidally dehiscent capsule; seeds usually reniform, smooth, sometimes with a funicule and strophiole, embryo peripheral, curved, perisperm hard, starchy, endosperm scanty or absent.

Distribution: Approximately 120 species in 13 genera, occurring mainly in the New and Old World tropics and subtropics, including a primary center of distribution in southern Africa; often weedy; 2 genera in the Guianas.

LITERATURE

Bogle, L.A. 1970. The genera of Molluginaceae and Aizoaceae in the southeastern United States. J. Arnold Arbor. 51: 431-462.

Burger, W.C. 1983. Flora Costaricensis, Aizoaceae. Fieldiana, Bot. ser. 2. 13: 213-217.

Eliasson, U.H. 1996. Molluginaceae. In G. Harling & L. Andersson, Flora of Ecuador 55: 3-11.

Endress, M.E. and V. Bittrich. 1993. Molluginaceae. In K. Kubitzki, The Families and Genera of Vascular Plants 2: 419-426.

[10] National Museum of Natural History, Department of Systematic Biology - Botany, NHB 166, Smithsonian Institution, Washington, D.C. 20013-7012, U.S.A.

Eyma, P.J. 1934. Aizoaceae. In A.A. Pulle, Flora of Suriname 1(1): 158-161.

Howard, R.A. 1988. Flora of the Lesser Antilles, Aizoaceae. 4: 195-200.

Lemée, A.M.V. 1955. Flore de la Guyane Française, Aizoacées, Mollugo. 1: 580.

Nevling, L.I. 1961. Aizoaceae. In R.E. Woodson & R.W. Schery, Flora of Panama 4: 422-427 = Ann. Missouri Bot. Gard. 48: 80-85.

Note: Regarding the MOLLUGINACEAE, Endress and Bittrich (1993) observed that "morphologically, the basic differences with the AIZOACEAE are in the structure of the androecium and the calyx, funicle length and the epidermis of leaves and stems." For additional information, see also note under family 30. AIZOACEAE.

KEY TO THE GENERA

1 Plants stellate-puberulent; sepals with linear apical appendage; style 1, stigmas 5 . . . . *1. Glinus*
  Plants glabrous; sepals not appendaged; styles and stigmas 3 . . . *2. Mollugo*

1. **GLINUS** L., Sp. Pl. 463. 1753.
   Type: G. lotoides L.

Annual herbs, stellate-pubescent, subsucculent; branches ascending to procumbent. Leaves alternate, opposite or whorled, linear to orbicular, often unequal in size at a node; stipules absent. Inflorescences axillary, fasciculate or seemingly cymose, cymes sometimes verticillate or flowers solitary; flowers sessile or pedicellate. Sepals 5, sometimes apically appendaged or cleft; petals absent; stamens 3-20, free, staminodes 0-20, often petaloid and divided at apex; ovary 3- to 5-locular, ovules numerous in each locule, style 1, short, stigmas 5. Capsule loculicidal, thin-walled or membranous; seeds small, strophiolate, testa smooth or granular.

Distribution: Approximately 10 species in the New and Old World (especially Africa) tropics and subtropics; weedy; 1 species in the Guianas.

1. **Glinus radiatus** (Ruiz & Pav.) Rohrb. in Mart., Fl. Bras. 14(2): 238, t. 55, f. 1. 1872. – *Mollugo radiata* Ruiz & Pav., Fl. Peruv. Chil. 1: 48. 1798. Type: not designated. – Fig. 29

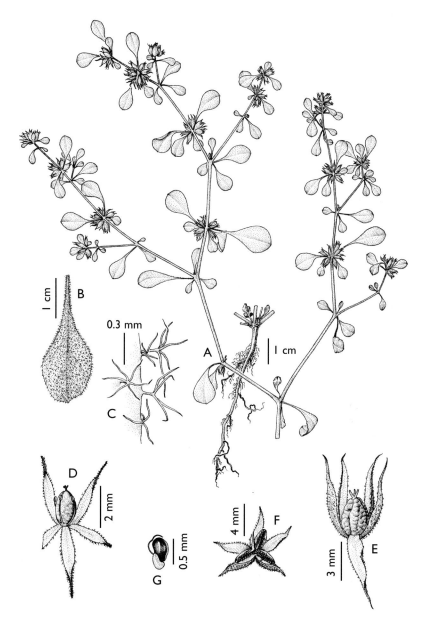

Fig. 29. *Glinus radiatus* (Ruiz & Pav.) Rohrb.: A, habit; B, leaf; C, stellate hairs; D, flower; E, fruit; F, dehisced fruit; G, seed. Drawing by E. Saavedra, reprinted from Nee, M. 1985. Flora de Veracruz 43.

Annual herb, stellate-puberulent; stems branched, prostrate or ascending, to 80 cm long. Leaves opposite or whorled, often unequal in size at a node; petiole 1-5(-12) mm long, with narrow hyaline wings; blade obovate, elliptical or spathulate, 10-40 x 2-15 mm, acute or obtuse at apex, cuneate to obtuse at base, narrowing gradually to base. Inflorescence an axillary fascicle of 3-8 flowers; pedicels 0-3 mm long, hairs branched and simple. Sepals 5, greenish, oblong or lanceolate, 2.5-4.5 x 0.8-1.5 mm, stellate-puberulent except near margin, with linear apical appendage of 0.5-1 mm long; stamens 3-5, filaments 1-1.5 mm long, anthers 0.5-0.7 mm long. Capsule ovoid or ellipsoid, 3-4 mm long, subtended by persistent perianth; seeds numerous, ovoid-reniform, 0.3-0.5 mm long, smooth, lustrous, pale reddish-brown, whitish strophiole filiform, shorter than to equalling the seed.

Distribution: New World tropics; in disturbed areas; 31 specimens seen, 3 from the Guianas (GU:2; FG: 1).

Selected specimens: Guyana: Rupununi Distr., Lethem, east bank of Takutu R., Irwin 773 (US); Rupununi Northern Savanna, Moreru Ranch, Goodland 1054 (US). French Guiana: 1 km en amont des Iles Yacarescin, Oyapock R., Oldeman 171h (US).

Note: The stellate hairs of the plant are shortly stalked, often with 6 rays of unequal length. The peculiar arillate apparatus emanating from the base of the seed consists of a filiform, white, funicle which follows the curve of the seed for about half-way around, as well as a minute (ca. 0.2 mm) but easily recognizable membranous, seemingly bilobed excrescence or somewhat inflated or scale-like, bilobed structure (strophiole), attached at the same place as the funicle.

2. **MOLLUGO** L., Sp. Pl. 89. 1753.
Lectotype: M. verticillata L.

Annual herbs, sometimes suffruticose, glabrous; branches often prostrate and spreading. Basal leaves crowded into a rosette, cauline leaves opposite or whorled; blade linear to spathulate, sometimes slightly fleshy; stipules small, scarious and caducous, or absent. Inflorescences of axillary fascicles, or solitary flowers; flowers pedicellate. Sepals 5, imbricate in bud, not apically appendaged; petals absent; stamens 3-10, usually 5, filaments filiform or widened, united at base; ovary 3- to 5-locular, ovules numerous, styles and stigmas as many as locules. Capsule loculicidal, membranous, after dehiscence with a central column bearing persistent funicles; seeds few to numerous, trigonous or more or less reniform, shiny, small, estrophiolate, testa granular or sculptured.

Distribution: Approximately 35 species in the New and Old World tropics and subtropics; weedy; 1 species in the Guianas.

1. **Mollugo verticillata** L., Sp. Pl. 89. 1753. Type: Herb. Linnaeus No. 112.4 (lectotype LINN, not seen) (designated Reveal *et al.*, Huntia 7: 212. 1987). – Fig. 30

Annual herb; stems often dichotomously branched, and prostrate-spreading, or ascending, to 30(-60) cm long. Leaves in whorls or pseudowhorls of 3-6 or more per node, sessile; basal blade oblanceolate, narrowly obovate or spathulate, cauline blade oblong-linear, spathulate-lanceolate or linear, 5-35 x 0.8-13 mm, apex obtuse to acute, tapering to base. Inflorescence an axillary fascicle or umbellule, of (1-)2-5(-18) flowers; pedicels filiform, 3-15 mm long, usually shorter than leaves. Sepals 5, ovate, oblong or elliptical, 1.8-3 x 0.5-1.8 mm, acute or obtuse, 3-veined, green on veins, with white margin; petals usually absent; stamens 3(4-5), to 1.5 mm long. Capsule ovoid, ellipsoid or reniform, 2.5-3 x 1.5 mm; seeds numerous, cochleate-reniform, 0.5-0.6 mm, flattened, dark to light brown or orange, longitudinally ridged, smooth or papillose.

Distribution: Cosmopolitan weed; 56 specimens seen, all from the Guianas (GU: 10; SU: 23 ; FG: 23).

Selected specimens: Guyana: St. Francis Mission, Mahaicony R., Davis 101 (K); Waranama Ranch, Intermediate Savannahs, Berbice R., Harrison *et al.* 1071 (K); Atkinson Field, East Bank Demerara, Harrison *et al.*1765 (K). Suriname: Tibiti Savanne, near village, Lanjouw & Lindeman 1818 (K); Lawa R., Versteeg 272 (U); Gonsoetoe-soela, Marowijne R., Geijskes 193 (U). French Guiana: Marouini R., Hoff 5244 (CAY); Awara, near carbet Tiouka, Lescure 635 (CAY); Jardin de Montjoly, E de Cayenne, Feuillet 579 (CAY).

Vernacular name: French Guiana: sulumonti (Galibi; translation: shrimp's moustache).

132

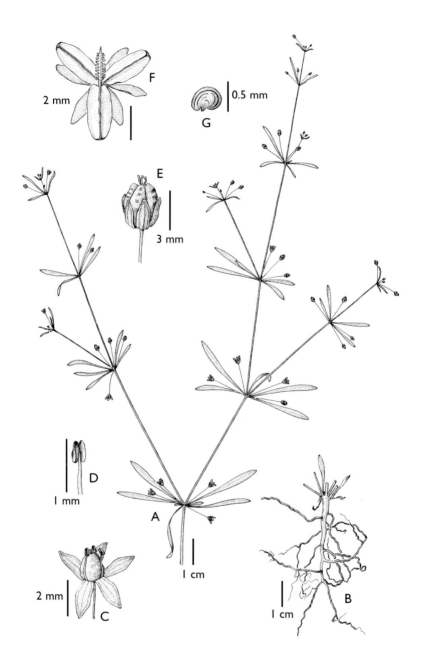

Fig. 30. *Mollugo verticillata* L.: A, habit; B, root system; C, flower; D, stamen; E, fruit; F, dehisced fruit; G, seed. Drawing by E. Saavedra, reprinted from Nee, M. 1985. Flora de Veracruz 43.

37. CARYOPHYLLACEAE

by

Robert A. DeFilipps and Shirley L. Maina[11]

Annual or perennial herbs, or subshrubs; stems often jointed and swollen at nodes. Leaves opposite and decussate, or whorled, simple, entire; exstipulate or less often with small, scarious stipules, often connate at base; often sessile or indistinctly petiolate, petiole often amplexicaul. Inflorescence a terminal, solitary flower, or a dichasial cyme or umbel. Flowers regular, bisexual, rarely unisexual; sepals 4-5, imbricate, free or united in a tubular, 4- or 5-toothed calyx, persistent; petals as many as sepals or sometimes some or all absent, free or nearly so, sometimes bifid and clawed; stamens in 1 or 2 whorls, sometimes in part absent, filaments free, basally united or epipetalous ones adnate to base of petal, anthers 2-locular, versatile, longitudinally dehiscent; ovary superior, sessile or shortly stipitate, 1(-5)-locular, ovules 1-numerous, campylotropous, placentae basal, central or free central, style 1, 1 lobed or 2-5 stigmas, or styles 2-5, free or united at base. Fruit a capsule, valvate or dehiscing by 2-5 apical teeth; seeds usually numerous, small, with mealy endosperm, embryo curved around perisperm.

Distribution: Approximately 2200 species in 86 genera, cosmopolitan, but most widely distributed in temperate regions of the northern hemisphere (with a center in the Mediterranean and Irano-Turanean regions), rarely indigenous to the tropics; in the Guianas 2 species in 2 genera.

LITERATURE

Bittrich, V. 1993. Caryophyllaceae. In K. Kubitzki, The Families and Genera of Vascular Plants 2: 206-236.

Duke, J.A. 1961. Caryophyllaceae. In R.E. Woodson & R.W. Schery, Flora of Panama 4: 432-448 = Ann. Missouri Bot. Gard. 48: 90-106.

Howard, R.A. 1988. Flora of the Lesser Antilles, Caryophyllaceae. 4: 210-212.

Van Ooststroom, S.J. 1934. Caryophyllaceae. In A.A. Pulle, Flora of Suriname 1(1): 150-153.

[11] National Museum of Natural History, Department of Systematic Biology - Botany, NHB 166, Smithsonian Institution, Washington, D.C. 20013-7012, U.S.A.

KEY TO THE GENERA/SPECIES

1 Leaves opposite, ovate-orbicular or subreniform; stipules ochraceous, often falling off rapidly; pedicels stipitate-glandular puberulent; sepals herbaceous with scarious margin; petals deeply bifid; seeds 1-12, tuberculate in lines .... 1. *Drymaria cordata*
   Leaves in whorls of ca. 15 leaves per node, linear, aristate; stipules silvery, persistent; pedicels puberulent, not glandular; sepals scarious; petals entire or slightly dentate; seeds 2-6, corrugated ..... 2. *Polycarpaea corymbosa*

1. **DRYMARIA** Schult. in Roem. & Schult., Syst. Veg. 5: 31. 1819. Lectotype: D. arenarioides Humb. & Bonpl. ex Schult.

Annual or perennial herbs; stems slender, usually prostrate, spreading, glabrous or pubescent, sometimes rooting at nodes. Leaves opposite or subverticillate; petiolate; stipules small, scarious, often persistent or falling off rapidly; blade glabrous to villous with often glandular hairs. Inflorescences few-flowered mono- or dichasial cymes; bracts scarious. Sepals 5, free, herbaceous with scarious margin; petals (0-3)5, white, often deeply bifid, sinus sometimes laciniate; stamens 2-5, filaments somewhat connate, flattened; ovary ovoid, sessile or substipitate, 1-locular, ovules few, campylotropous on free central placenta, 3 styles united at base. Fruit an ovoid, 3-valved capsule; seeds 1-numerous, globose-reniform, cochleate or hippocrepiform, usually granular or tuberculate, rarely smooth.

Distribution: Approximately 48 species, the great majority in tropical and subtropical America; 1 species in the Guianas.

Literature: Duke, J.A. 1961. Preliminary revision of the genus Drymaria. Ann. Missouri Bot. Gard. 48: 173-268.

1. **Drymaria cordata** (L.) Schult. in Roem. & Schult., Syst. Veg. 5: 406. 1819. – *Holosteum cordatum* L., Sp. Pl. 88. 1753. Type: not designated (according to Howard 1988: 211, Herb. Linnaeus No. 109.1 is post-1753). – Fig. 31

Annual herb to 45 cm long, erect to prostrate; stems branched, angled, often rooting at nodes, glabrous to densely glandular-puberulent. Petiole filiform, 2-15 mm long; stipules membranous, multilacerate, to 2 mm long; blade orbicular, ovate-orbicular or reniform, 5-25 x 5-30 mm, obtuse or acute and mucronulate at apex, round to cordate at base, glabrous or puberulent. Inflorescences terminal or axillary, few-flowered dichasial cymes, lax; pedicels 2-15 mm long, densely stipitate-glandular

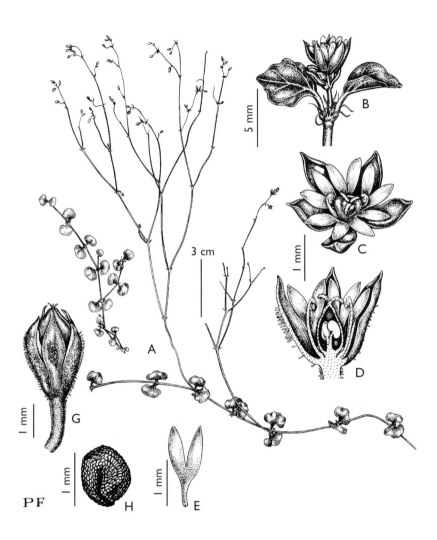

Fig. 31. *Drymaria cordata* (L.) Schult.: A, on left, young branchlet, on right, fruiting branch, x 2/3; B, inflorescence, x 4; C, flower, from above, x 16; D, flower longitudinally dissected, x 16; E, petal, x 16; G, fruit, x 12; H, seed, x 18. Drawing by P. Fawcett, reprinted from Correll, D.S. & H.B. Correll. 1982. Flora of the Bahama Archipelago.

puberulent. Sepals 5, ovate, oblong-lanceolate or lanceolate, 2.5-3.5 mm long, glabrous except for stipitate-glandular puberulent midvein; petals 5, white, deeply clawed and bifid, shorter than calyx, 2-3 mm long, lobes linear; stamens 2-3(-5), filaments ca. 2 mm long; ovary ovoid, styles 3, 0.5-1 mm long. Capsule ovoid, slightly shorter than calyx, 1.5-2.5 mm long, shortly pedicellate; seeds 1-12, cochleate, 1.0-1.5 mm long, dark reddish-brown or black, densely tuberculate in lines.

Distribution: Widely distributed in the tropics and subtropics of Old and New Worlds; disturbed areas; 24 collections studied, all from the Guianas (GU: 4; SU: 10; FG: 10).

Selected specimens: Guyana: Vicinity of Mazaruni Forest Station, Maguire & Fanshawe 23562 (MO, U); Mazaruni R. Settlement, Mell & Mell 232 (NY); Potaro-Siparuni Region, Kaieteur Falls National Park, Hahn 4472 (U, US). Suriname: Paramaribo, Cultuurtuin, Lindeman 5722 (U), van Doesburg 27 (U); Paramaribo, Fernandesweg, Reijenga 624 (U). French Guiana: Village of Saül, Mori et al. 23381 (NY); Cayenne, rue Justin Catayee, Ducatillon & Gelly 83 (CAY); Cayenne, Chez Françoise, Alexandre 383 (CAY).

Use: The plant is used in a medicinal bath for children in French Guiana.

Vernacular names: Suriname: piki fowroesopo. French Guiana: mignonette, petit quinine, timignonette (Creole).

2. **POLYCARPAEA** Lam., J. Hist. Nat. 2: 3, 5. 1792, nom. cons.
   Type: P. teneriffae Lam., typ. cons.

Annual, biennial or perennial herbs, rarely suffruticose; stems usually erect, branched, glabrous or pubescent. Leaves opposite or whorled; sessile at swollen nodes; stipules conspicuous, scarious; blade usually linear-subulate or lanceolate, rarely elliptical or spathulate. Inflorescences terminal dichasial cymes, often contracted or capitate; bracts scarious. Sepals 4-5, free, scarious or rarely herbaceous with scarious margin; petals 4-5, entire or dentate, shorter than sepals; fertile stamens 4-5, staminodes 4-5, filaments free or connate at base; ovary ovoid, sessile or shortly stipitate, 1-locular, ovules few, campylotropous on basal placenta, style 3-fid, 3-dentate, or capitate and 3-sulcate. Fruit a 3-valved capsule, ovoid or ellipsoid; seeds 1-few, cochleate, oblong or reniform, smooth or tuberculate.

Distribution: Approximately 30 species in tropical and subtropical areas of the Old and New Worlds; 1 species in the Guianas.

Literature: Lakela, O. 1962. Occurrence of species of Polycarpaea Lam. (Caryophyllaceae) in North America. Rhodora 64: 179-182.

1. **Polycarpaea corymbosa** (L.) Lam., Tabl. Encycl. Méth., Bot. 2: 129. 1797. – *Achyranthes corymbosa* L., Sp. Pl. 205. 1753. Type: not designated. – Fig. 32

Polycarpaea atherophora Steud., Flora 26: 763. 1843. Type: Suriname: In arenosis aridis, 1842, Hostmann & Kappler 598 (K, MO, U).

Annual, rarely perennial, herb; stems erect, 5- ca. 30 cm tall, terete, often highly branched, flocculose-tomentose above. Leaves in whorls of up to ca. 15(-26) leaves per node; stipules silvery-white, scarious, aristate, multilacerate, 2-5 mm long, persistent; blade linear or linear-subulate, 3-15 x 0.2-0.3 mm, long-aristate with a distinctly differentiated apical arista, glabrous or sparsely pilose, sometimes mucilaginous. Inflorescence a many-flowered, dichasial cyme; pedicels 2-5 mm long, puberulent. Sepals usually 5, silvery-white, lanceolate, 2.5-3.5 mm long, acute to aristate, scarious; petals 5, sordid-white, entire or slightly dentate, obovate or ovate-oblong, 0.5-1 mm long, obtuse or denticulate, much shorter than sepals; stamens 5, filaments flattened, 0.5-1 mm long; ovary obovoid or ovoid-oblong, shortly stipitate, style to 0.5 mm long, 3-sulcate at apex. Capsule ellipsoid or obovoid, 1-2.5 mm long; seeds 2-6, cochleate, ca. 0.5 mm wide, minutely corrugated, pale brown.

Distribution: Tropical and subtropical regions of the Old and New Worlds; weeds, often from sandy habitats; 23 collections studied, all from the Guianas (GU: 12; SU: 11).

Selected specimens: Guyana: Vicinity of Karanambo, Maas *et al.* 7693 (CAY, US); Savanna near Oreala, Pulle 533 (U); Rupununi Northern Savanna, 3 mi S of Meritiziero, Goodland 537 (US). Suriname: Zanderij I, Archer 2837 (U, US); Para Distr., Lobin Savanne, Teunissen *et al.* LBB 14521 (U); Tibiti Savanne, Lanjouw & Lindeman 1663 (U).

Note: Several specimens from Guyana and Suriname have been annotated as var. *brasiliensis* (Cambess.) Chodat & Hassl. by Olga Lakela; this variety, which occurs in Brazil and Paraguay, is said to be a perennial with woody napiform root and strongly revolute leaves, whereas typical *P. corymbosa* is an annual with fibrous roots from a slender main root and flat or slightly revolute leaves. All specimens seen for this treatment conform to typical *P. corymbosa*. For a brief discussion of variation in this species, see Lakela (1962).

138

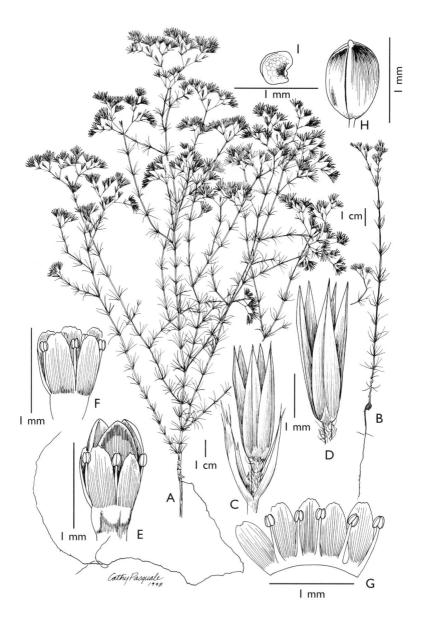

Fig. 32. *Polycarpaea corymbosa* (L.) Lam.: A, habit of branched plant; B, habit of unbranched plant; C, bracteate flower with sepals; D, flower with sepals, the bracts removed; E & F, flower with petals, the sepals removed; G, stamens and petals; H, capsule; I, seed. (A, C-I, from Pulle 533; B, from Maas *et al.* 3770). Drawing by Cathy Pasquale.

54. SARRACENIACEAE
by
ANDRÉ S.J. VAN PROOSDIJ[12]

Perennial herbs, sometimes climbing. Stem simple or branched. Leaves rosulate or alternate, not stipulate, sessile; phyllodes present or absent; mature leaves transformed into large, tubular pitchers (insect traps) with a reduced blade, indument inside pitcher differentiated (see note to *Heliamphora*); primary veins in the pitcher parallel, secondary veins reticulate, veins in the lid reticulate. Inflorescences simple, terminal, single-flowered or racemose; bracts absent or present, then scale-like or foliaceous, sessile; bracteoles absent or present. Flowers bisexual, regular; sepals 4-6, 1-cyclic; petals absent or 5, 1-cyclic, free, strongly imbricate; stamens 10-numerous, in 1 whorl, or numerous and originating from an irregular primordial plexus, free, anthers oblong-linear, 2-locular, 4-thecate, basifixed, sometimes during anthesis reflexing from introrse to extrorse, or anthers dorsifixed, introrse and not reflexing, laterally dehiscing or dehiscing by caudal appendages; pollen 20-35 x 20-30(-46) mm, (3-)4-9-colporate, exine verruculate or minutely scrobiculate; staminodes absent; ovary superior, 3(-4)-5-locular, glabrous, pubescent or tuberculate, ovules numerous, axial, anatropous, style 1, simple, erect, shallowly lobed, 5-branched or expanded into broad umbrella-shaped disc with 5 stigmas. Fruits loculicidal capsules; seeds many, ovoid, obovoid or clavate, compressed, fimbriate or basally tailed, embryo small or relatively large, erect, embedded in (copious) endosperm, cotyledons 2, short.

Distribution: A Neotropical family of 3 genera and 15 species; *Sarracenia* L. from the E coast of the United States (8 species), the monotypic genus *Darlingtonia* Torr. from N California and adjacent Oregon and *Heliamphora* Benth. with 6 species restricted to the Guayana Highlands; in the Guianas only *Heliamphora nutans* Benth. is present.

Ecology: In wet, sandy or boggy places, deficient in nitrogenous nutrients.

---

[12] Nationaal Herbarium Nederland, Utrecht University branch, Heidelberglaan 2, 3584 CS Utrecht, The Netherlands.

140

LITERATURE

Bentham, G. 1840. On the Heliamphora nutans, a new Pitcher-plant from British Guiana. Trans. Linn. Soc. London 18: 429-433.

Gleason, H.A. *et al.* 1931. Botanical results of the Tyler-Duida Expedition. Bull. Torrey Bot. Club 58: 366-368.

Maguire, B. 1970. On the Flora of the Guayana Highland. Biotropica 2: 85-100.

Maguire, B. 1978. Sarraceniaceae. In B. Maguire *et al.*, The Botany of the Guayana Highland Part 10. Mem. New York Bot. Gard. 29: 36-62.

Steyermark, J.A. 1951. Sarraceniaceae. In J. A. Steyermark *et al.*, Botanical Exploration in Venezuela 1. Fieldiana, Bot. 28: 239-242.

Tracey, S.M. & F.S. Earle. 1896. Mississippi Fungi. In Mississippi Agric. Exp. Sta. Bull. 34: 136-153.

Uphof, J.C.Th. 1936. Sarraceniaceae. In A.H.G. Engler & K.A.E. Prantl, Die natürlichen Pflanzenfamilien ed. 2. 17b: 704-727.

1. **HELIAMPHORA** Benth., Proc. Linn. Soc. Lond. 1: 53. 1840.
Type: H. nutans Benth.

Phyllodes present, often conspicuous; pitchers ventricose or essentially tubular, up to 50 cm long, outside with scattered bifid hairs and small glands, base attenuate, ventral wings paired, apical leaflet small (absent in *H. macdonaldae* Gleason), cup-shaped with narrow base, not covering the pitcher mouth, inside producing nectar. Inflorescences racemes; peduncle exceeding leaves; bracts foliaceous; bracteoles absent; pedicels slender and drooping or stout and erect, glabrous to densely pubescent. Flowers nodding; sepals 4(-6), petaloid, white to red, glabrous; petals absent; stamens 10-numerous, 1-cyclic, filaments stout, thick at base, anthers basifixed, sagittate, during anthesis reflexing from introrse to extrorse, dehiscing through upwardly directed, caudal appendages; pollen subglobose, ca. 30-35 x 20-25 mm, (3-)4-5-colporate, exine verruculate; ovary 3(-4)-locular, densely covered with bifid hairs, style slender, cylindrical, glabrous, stigma shallowly truncately 3(-4)-lobed. Capsules splitting downwards in three parts, held together by the style; seeds ovoid, glabrous, irregularly scario-fimbriately winged, embryo small, embedded in copious endosperm. Chromosome number 2n = 42 (Maguire, 1978).

Distribution: Species 6, in the Guayana Highlands (Roraima Formation of southern Venezuela, contiguous Guyana and Brazil); on higher altitudes in marshy or wet savanna-like places, on the so-called tepuis (abruptively arising high sandstone hills with rugged summits); in Guyana only one species from Mt. Roraima.

N o t e s : The number of species is not certain. Maguire (1978) stated that "several populations of possible narrow genetic differences are held intact because of geographic isolation". Further research is needed to answer the questions about the number of species and their limits.

The indument on the inner side of the pitcher is highly specialised. The upper part of the pitcher is covered with many strong, long, downwardly directed, non-secreting hairs (*H. macdonaldae* Gleason glabrous), the middle part is glabrous and the lower part covered with scattered, smaller, rougher, reflexed hairs. The hairs from the lowest part have the appearance of secreting hairs.

Many authors have already mentioned the carnivorous function of the pitchers.

Living creatures have been found inside the pitchers, such as bacteria, amoebae and insects. Most of the data are from *Sarracenia* species, hardly anything is known from *Heliamphora* species. Only Uphof (1936) mentioned flies and butterflies in *H. tyleri* Gleason.

Parasitic fungi were found on *Sarracenia* species (Tracey & Earle, 1896; Uphof, 1936), but no data are yet known from *Heliamphora* species.

Halfway the pitcher, just above the basal part, a small pore in the seam is present. Through this pore the plant can get rid of the water surplus from heavy rainfall. Downward directed hairs as present at the inside of the pitcher prevent insects from escaping the pitcher through the pore. The pore is difficult to see, because it is situated between the lateral wings. As a result of the pressing of the plant, this pore is usually not visible in herbarium specimens.

1. **Heliamphora nutans** Benth., Proc. Linn. Soc. Lond. 1: 53. 1840. Type: Venezuela, Mt. Roraima: Ro. Schomburgk ser. I, 1050 (holotype K). – Fig. 33

Stem short, simple. Leaves rosulate; phyllodes linear to elliptic, 3.0-4.5 x 0.5 cm, laxly pubescent, pitchers conspicuously ventricose, prominently constricted above and below middle, (8-)12-18(-32) cm long, green, partly dotted purplish brown or red, ventral wings 1-4 mm high, in lower parts up to 10 mm, apical leaflet 6-8(-10) x 6-8(-10) mm, red or purple-red, upper hairy zone inside covering one-third of the pitcher length. Inflorescence (2-)3-4(-6)-flowered; ebracteate or bracteate at 1/4 to 1/3 of peduncle length; peduncle 30-50 cm long, glabrous; pedicel slender, drooping, 2.5-3.5(-4.0) cm long, glabrous; bracts 4-8 x 2-3 cm, bracteoles 1.5-2.5 x 1.0-2.0 cm; bracteole from

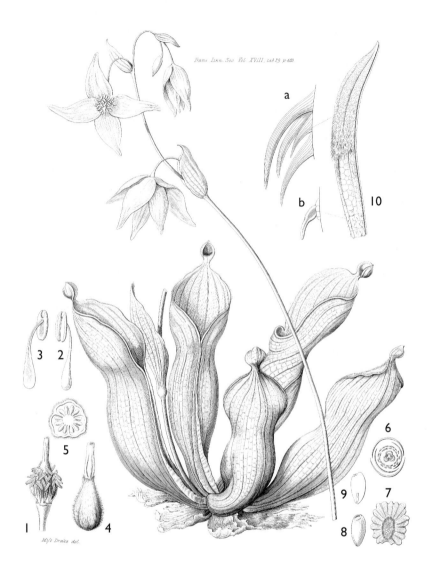

Fig. 33. *Heliamphora nutans* Benth.: habit with inflorescence; 1, flower, sepals removed ; 2-3, stamen; 4, pistil; 5, transverse section of ovary; 6, diagram of flower; 7, seed; 8, seed, testa removed; 9, longitudinal section of seed; 10, detail of leaf innerside. Reprinted from Bentham, G. 1840. Trans. Linn. Soc. London 18: t. 29.

first flower up to 4 x 3 cm, shorter than pedicel, glabrous. Flowers: sepals 4-5, white turning pink or red, ovate-elliptic, acuminate, 3.0-4.5 x 1.3-2.0 cm, glabrous; stamens 21-27(-32), filaments 5-6 mm long, glabrous, anthers 2.5-3.5 mm long, glabrous; ovary densely covered with bifid hairs. Seeds 1.5-2.0 x 0.8-1.0 mm, glabrous, wings ca 1 mm wide, testa brown.

Distribution: In the Roraima chain: Serra do Sol, Roraima, Ilu-tepui axis and the nearby Ptari-tepui and Chimanta-tepui complexes of the Gran Sabana region of eastern Guayana, in this region found from Mt. Roraima and further north-west to Ilu-tepui; occasional to frequent in clumps among dense mats of low vegetation, open shrubs, low forest in marshy places on white sand and boulders, at 1600-2300 m alt.; in Guyana only on Mt. Roraima (GU: 8).

Selected specimens: Guyana: Mt. Roraima: Ri. Schomburgk 983; Abbensetts 3, 42; top of mountain, Tate 363; Our House, ledge & top, Im Thurn 258; N slope, Renz 14202 (U), 14267 (U); 1 km N of tip of northern "prow", El Dorado swamp, Hahn *et al.* 5528.

Phenology: Flowering in Jan., Feb., Nov. and Dec. Fruiting in Feb., Nov. and Dec.

Note: Steyermark (1951) mentioned the similarity between *H. nutans* and *H. minor* Gleason. He stated that "the differences of pubescence of pedicel and smaller size of sepals separating *H. minor* from *H. nutans* have been found to be variable and inconsistent characters". Maguire (1978) regards them still as distinct species.

55. DROSERACEAE

by

RODRIGO DUNO DE STEFANO[13]

Annual or perennial insectivorous herbs. Roots adventitious (some extra-neotropical species with tubers). Stems 0.1-30 cm, aerial (subterranean in some extra-neotropical species). Leaves generally stipulate, alternate, simple (variously lobed or peltate in some extra-neotropical species), covered with glandular hairs; stipules intrapetiolar (lateral or absent in some extra-Guianan species); petiole grading into blade or well defined, hairy or glabrous; blade membranaceous, linear to circular, margin covered with glandular hairs. Inflorescences terminal, rarely lateral, unbranched, one-sided or dichasial cymes, sometimes flowers solitary; peduncle slender, up to 20 cm long, with glandular hairs, non-glandular hairs or glabrous (also in bracts, pedicels and sepals); bracts minute, green, red or scarious; pedicels present. Flowers generally 5-merous, actinomorphic (zygomorphic in some extra-tropical species), hermaphrodite; sepals joined at base or free; petals free, white or pink, rarely yellow, membranaceous, spathulate or obovate, usually persistent; stamens generally 5, filaments filiform, terete, anthers usually bilocular; pollen generally released in tetrads, generally spinulose; ovary superior, glabrous, carpels generally 3, united, placentation parietal, styles 3-5, usually partite at base or unbranched, rarely united, often further dividing distally, stigmas tapered or swollen, often papillose. Fruits dehiscent (rarely indehiscent in extra-Neotropical species), papery capsules; seeds numerous (few in some extra-Neotropical species), embryo embedded in copious endosperm, rich in starch, oil and protein, cotyledons minute.

Distribution: Cosmopolitan with centre of diversity in Australia, 4 genera and ca. 120 species, the genus *Drosera* is the only one in the Neotropics.

[13] Fundación Instituto Botánico de Venezuela, Apartado 2156, Caracas 1010-A, Venezuela. I would like to thank for the support and kind collaboration of the Real Jardín Botánico de Madrid, its director Dr. M.T. Tellerias, my supervisor Dr. S. Castroviejo and the curator Dr. M. Velayos. In Utrecht, to Dr. A.R.A. Görts-van Rijn who carefully checked the manuscript. The botanical material was provided by the following herbaria: AAU, B, BM, BRG, CAY, F, K, MA, MO, NY, P, PORT, U, US, VEN.

Ecology: The Droseraceae is a family of insectivorous (carnivorous syndrome) herbs; commonly growing in bogs and other waterlogged soils; the carnivorous habit may be an evolutionary response to their growth in media containing little nitrogen.

Economic uses: *Aldrovandra, Dionaea, Drosophyllum* and several *Drosera*'s from the Old World are grown, usually in greenhouses as curiosities and as ornamental plants.

LITERATURE

Brummer-Dinger, C.H. 1955. Notes on Guiana Droseraceae. Acta Bot. Neerl. 4: 136-138.

De Candolle, A.P. 1824. Droseraceae. Prodromus Systematis Naturalis Regni Vegetabilis I: 317-320.

Diels, L. 1906. Droseraceae. In H.G.A. Engler, Das Pflanzenreich IV. 112. (Heft 26): 1-136.

Duno de Stefano, R. & A. Culham. 1995. Dos especies nuevas del género Drosera (Droseraceae) en Venezuzela y otros comentarios taxonómicos. Novon 5: 240-245.

Eichler, A.G. 1872. Droseraceae. In C.F.P. von Martius, Flora Brasiliensis 14(2): 385-398, t. 90-91.

Fernández-Pérez, A. 1965. Plantas Insectívoras. II. Droseraceas de Colombia. Caldasia 9: 219-232.

Kunth, C.S. 1821. Droseraceae. In F.W.H.A. von Humboldt, A.J.A. Bonpland & C.S. Kunth, Nova Genera et Species Plantarum ed. qu. 5: 390-391.

Maguire, B. & J.J. Wurdack. 1957. The Botany of the Guayana Highland - Part II. Mem. New York Bot. Gard. 9: 331-336.

Planchon, J.E. 1848. Sur la Famille des Droséracées. Ann. Sci. Nat., Bot. ser. 3. 9: 79-99; 185-207; 285-309.

Saint-Hilaire, A.F.C.P. de. 1829. Droseraceae. Flora Brasiliae Meridionalis 2: 93-96.

Santos, E. 1989. Droseráceas. In P.R. Reitz, Flora Ilustrada Catarinense. 23 pp.

Seine, R. & W. Barthlott. 1994. Some proposals on the infrageneric classification of Drosera L. Taxon 43: 583-589.

Wynne, F.E. 1944. Drosera in Eastern North America. Bull. Torrey Bot. Club 71: 166-174.

1. **DROSERA** L. Sp. Pl. 281. 1753.
Type: D. rotundifolia L.

Perennial or annual (not in Guianan species) insectivorous herbs; acaulescent or caulescent. Stems 0.1-10 cm long, aerial. Leaves generally basal, stipules (absent in some extra-Guianan species) intrapetiolar, 4-7-partite; petiole grading into blade or well defined, hairy or glabrous; blade obovate to circulate (linear in some extra-Guianan species), hairy or glabrous, margin bearing glandular hairs. Inflorescences generally one-sided cymes, (1)-6-15-(25)-flowered, sometimes flowers solitary; peduncle up to 20 cm long, slender, erect or curved near base, with glandular hairs, non-glandular hairs or glabrous; bracts along peduncle and sometimes one by flowers. Flowers perfect, open for a short time only, sometimes cleistogamous; sepals ovate or narrowly ovate, free or joined at base, margin entire or glandular, with glandular hairs, non-glandular hairs or glabrous without; petals free, white or pink, spathulate or obovate, alternate with stamens, margin entire; ovary with 3 styles, bipartite at base, rarely 5 and unbranched. Fruits dehiscent papery capsules; seeds numerous, minute, foveolate, foveolate-reticulate, reticulate or papillose, narrowly oblongoid, obovoid to circulate.

Distribution: Cosmopolitan, 120 species; about 25 species in the Neotropics, 8 in the Guianas.

Ecology: Plants growing in wet sand savanna, bogs, scrub vegetation, sandy areas along streams and coastal vegetation; between 10-2700 m alt.

Subdivision: Seine & Barthlott (1994) published an infrageneric classification of *Drosera*, based on the previous one of Diels (1906) and Planchon (1848). They recognized three subgenera, the largest one is subgenus *Drosera*, the only one present in the Neotropics. This subgenus is divided into eleven sections, but only three are present in the Neotropics. The cosmopolitan section *Drosera* includes almost all the neotropical species, except *D. sessilifolia* in the section *Thelocalyx* and *D. meristocaulis* Maguire & Wurdack in the monotypic section *Meristocaulis* (endemic to Cerro de La Neblina, Venezuelan-Brazilian border).

Vernacular name: sundew (English).

KEY TO THE SPECIES

1 Leaves cuneate with an indistinct petiole; styles 5, not bipartite (section *Thelocalyx*) .... 8. *D. sessilifolia*
Leaves spathulate, petiole usually distinct and narrow; styles 3, bipartite from the base (section *Drosera*) 2

2 Caulescent herb, stem up to 10 cm long, covered by marcescent leaves; stipules 3-7 mm long, coriaceous and conspicuous; petioles very hairy; peduncles and sepals with glandular hairs and sessile glands, rarely glabrous .... 7. *D. roraimae*
Acaulescent herbs, rarely caulescent and the stem very short, 1-3 cm long; stipules 1-5 mm long, membranaceous and almost inconspicuos; petioles glabrous or hairy; peduncles and sepals with long non-glandular hairs, glandular hairs and sessile glands, sometimes glabrous 3

3 Leaf blades usually > 2 times as long as wide; petiole 4-40 mm long; inflorescences (2-)10-30 cm long; seeds papillose 4
Leaf blades < 2 times as long as wide; petiole 3-15 mm long; inflorescences 1-7(-10) cm long; seeds never papillose 5

4 Petals pink; seeds 0.35-0.4 mm long, with papillae arranged in more or less longitudinal rows .... 2. *D. capillaris*
Petals white; seeds 0.25-0.3 mm long, with papillae arranged irregularly ... . 5. *D. intermedia*

5 Peduncles and sepals with long non-glandular hairs and smaller glandular hairs; sepals reflexed at anthesis 6
Peduncles with glandular hairs or sessile glands; sepals erect at anthesis ... 7

6 Blades widely obovate, 4-7 x 2.5-5 mm; sepals free, narrowly ovate and marcescent; fruits after ripening not cup-like .... 3. *D. cayennensis*
Blades broadly to very widely obovate, 2.5-5 x 2.5-4 mm; sepals joined at base, narrowly ovate and deciduous; fruits after ripening cup-like .... 6. *D. kaieteurensis*

7 Petioles 3 mm long; inflorescences and sepals glabrous, peduncles 1.5-2 cm long, 1-3-flowered .... 1. *D. biflora*
Petioles 5-6 mm long; inflorescences and sepals with sessile glandular hairs, peduncles 3-8 cm long, (1-)3-10-flowered .... 4. *D. esmeraldae*

1. **Drosera biflora** Humb. & Bonpl. ex Roem. & Schult., Syst. Veg. 6: 763. 1820. Type: Venezuela, Amazonas, Atabapo, Humboldt & Bonpland 1100 (holotype B).

Drosera pusilla Kunth in Humb., Bonpl. & Kunth, Nov. Gen. Sp. ed. qu. 5: 390, t. 490, f.1. 1821. Type: Venezuela, Amazonas, Atabapo, Humboldt & Bonpland 1100 (holotype P).

Acaulescent glabrous herb forming a rosette, no more than 2 cm tall. Leaves spathulate, 6 mm long, petiole and blade well defined; stipules 1-1.5 mm long, 4-5-partite; petiole 3 x 0.8 mm; blade circulate or broadly elliptic, 3 x 2 mm. Inflorescences 1-2 per plant, 2-3 cm long, 2-3-flowered; peduncle erect, 1.5-2 cm long. Flowers with sepals ovate, joined at base, 2-3 x 1.5 mm, margin entire; petals white, spathulate, 3-4 mm long; stamens 2.5-3 mm long; pollen unknown; ovary with 3 styles, bipartite near base. Seeds foveolate, ovate, almost circulate, 0.35-0.4 x 0.3 mm, covered with granulate wax.

Distribution: Venezuela (Amazonas), Guyana and Brazil (Maranhão); wet and sandy savanna, shrub savanna, between 100-200 m alt.; 7 collections studied, 2 from the Kaieteur Savanna in Guyana (GU: 2).

Specimens examined: Guyana: Kaieteur Savanna, Jenman 912, 1293 (BRG).

Note: *D. pusilla* has been used as the correct name for this species since the publication of the Nova genera et species plantarum (Kunth 1815-1825). In this work, *D. biflora* is mentioned as a synonym. However, Roemer & Schultes validly published *D. biflora* one year earlier, using the name proposed by Willdenow on the sample deposited in Berlin.

2. **Drosera capillaris** Poir. in Lam., Encycl. 6: 299. 1804. Type: not designated.

Drosera tenella Humb. & Bonpl. ex Roem. & Schult. Syst. Veg. 6: 763. 1820. Type: Venezuela, Sucre, La Cuchilla, between Guanaguana and Caripe, Humboldt & Bonpland 216 (holotype B, not seen, isotype P).

Acaulescent herb forming a rosette, no more than 3 cm tall. Leaves spathulate, 7-30 mm long, petiole and blade well defined; stipules 3.5-5 mm long, 5-7-partite; petiole 4-20 x 0.5-1 mm, hairy beneath; blade obovate, 3-10 x 1.5 mm, glabrous or hairy beneath. Inflorescences 1-5 per plant, 2-22 cm long, 2-9-flowered; peduncle erect or curved near base, 1-20 cm long, glabrous or with few sessile glands. Flowers with sepals ovate, joined at base, 1.5-4 x 0.5-1.5 mm, margin entire, glabrous or with few sessile glands outside; petals pink, spathulate, 2.5-4 x 0.5-1 mm; stamens 3-3.5 mm long; pollen tetrad 60-70 μm diameter, spines and spinules well defined, spine 1.5 μm long, 12-18 μm^2; ovary with 3 styles, bipartite near base. Seeds papillose in more or less longitudinal rows, ellipsoid, 0.35-0.4 x 0.15-0.2 mm.

Distribution: Throughout the Neotropics; wet savanna and bogs, between 10-2000 m alt.; 50 collections studied, 30 from the Guianas (GU: 2; SU: 23; FG: 5).

Selected specimens: Guyana: Gunn's, Essequibo R., Jansen-Jacobs *et al.* 1392 (K, U); Jenman 3769 (K). Suriname: Distr. Saramacca, Kappel Savanna, Tafelberg, van Donselaar 2938 (U); savanna near Matta, Jansma 25 (U). French Guiana: 7 km W of Kourou, van Donselaar 2595 (U); Savane de Kourou, Sastre 1322 (P).

Vernacular names: yeberu, yeberubina (Arow.)

3. **Drosera cayennensis** Sagot ex Diels in Engl., Pflanzenr. IV. 112 (Heft 26): 86. 1906. Type: French Guiana, Sagot 1228 (holotype P, not seen).

Drosera sanariapoana Steyerm., Fieldiana, Bot. 28: 243. 1952. Type: Venezuela, Amazonas, vicinity of Sanariapo, near Río Sanariapo, tributary of Río Orinoco, Steyermark 58472 (holotype NY, isotype VEN).
Drosera colombiana A. Fernández, Caldasia 9: 226, f. 1. 1965. Type: Colombia, Meta, Llanos de San Martín, Dryander 3019 (holotype US, isotype COL).
Drosera panamensis M.D. Correa & A.S. Taylor, Ann. Missouri Bot. Gard. 63: 390. 1977. Type: Panama, La Yeguada, altos de Baltazar y el Veladero, Correa *et al.* 2215 (holotype PMA, not seen, isotypes COL, F, K, MO, U, US).

Acaulescent herb forming a rosette, no more than 2 cm tall. Leaves spathulate, 7-15 mm long, glabrous, petiole and blade well defined; stipules 1-2.5 mm long, 4-6-partite; petiole 3-7 x 0.8 mm; blade widely obovate, 4-7 x 2-5 mm. Inflorescences 1-2 per plant, 5-9 cm long, 1-4(-7)-flowered; peduncle erect, 3-8 cm long, with long non-glandular hairs and smaller glandular hairs. Flowers with sepals narrowly ovate, free from base, 3-6 x 0.8-1.2 mm, margin entire, with long non-glandular hairs and smaller glandular hairs outside; petals pink, spathulate, 6 x 1.5-2 mm; stamens 4-5 mm long; pollen unknown; ovary with 3 styles, bipartite near base. Seeds foveolate (immature), 0.3-0.45 x 0.14-0.16 mm, covered with granular wax.

Distribution: Panama, Colombia, Venezuela, Suriname and French Guiana, expected in Guyana; wet savanna and littoral bogs, common in the littoral area, 100-600 m alt.; 20 collections studied, 8 from the Guianas (SU: 6; FG: 2).

Selected specimens: Suriname: Gros Savanna km 103, van Donselaar 739 (U); Distr. Saramacca, Kappel Savanna, S of Tafelberg, Kramer & Hekking 3041, 3260, 3280 (U). French Guiana: Leprieur 145 (M).

4. **Drosera esmeraldae** (Steyerm.) Maguire & Wurdack, Mem. New York Bot. Gard. 9: 335. 1957. – *Drosera tenella* Humb. & Bonpl. ex Roem. & Schult. var. *esmeraldae* Steyerm., Fieldiana, Bot. 28: 244. 1952. Type: Venezuela, Amazonas, between Esmeralda Savanna and SE base of Cerro Duida, Steyermark 57850, (holotype F, not seen, isotype NY, US).

Acaulescent herb forming a rosette, no more than 2 cm tall. Leaves spathulate, 5-12 mm long, petiole and blade well defined; stipules 2-2.5 mm long, 4-5-partite; petiole 5-6 x 0.8-1 mm, hairy beneath; blade broadly obovate to circulate, 2-3 x 2-3 mm, glabrous or hairy beneath. Inflorescences 1-2-(-4) per plant, 3.5-10 cm long, (1-)3-6-flowered; peduncle erect, 3-6.5 cm long, glabrous or with few glandular sessile hairs. Flowers with sepals ovate, joined at base, 2.5 x 0.8 mm, margin entire, with glandular hairs outside; petals white or pink, spathulate, 3-3.5 mm long; stamens 3 mm long; pollen tetrad 90-130 μm diameter, spines and spinules well defined, spines 3.5 μm long, 6-14 μm²; ovary with 3 styles, bipartite near base. Seeds foveolate 0.30-0.35 x 0.15-0.2 mm, covered with granulate wax.

Distribution: Colombia (Amazonas and Vaupés), Venezuela and Guyana; wet savanna, between 10-400 m alt.; 13 collections studied, 1 collection from Guyana (GU: 1).

Specimen examined: Guyana: savanna near Kuyuwini Landing, Jansen-Jacobs *et al.* 3157 (U).

5. **Drosera intermedia** Hayne, J. Bot. (Schrader) 1: 37. 1800. Type: [Germany], Hamburg near Eppendorf, Hayne s.n. (holotype H, not seen).

Acaulescent or very short stemmend herb forming a rosette, no more than 3 cm tall. Leaves spathulate, 20-40 mm long, petiole and blade well defined; stipules 2-4 mm long, 3-7-partite; petiole 15-30 x 0.5 mm, hairy beneath; blade obovate, 5-10 x 1.5-3 mm, glabrous or hairy beneath. Inflorescences 1-4 per plant, 5-20 cm long, 3-15-flowered; peduncle erect or curved near base, 3-15 cm long. Flowers with sepals ovate, joined at base or free, 2.5-3 x 1 mm, margin entire; petals white, spathulate, 3-6 x 2-3 mm; stamens 2-5 mm long; pollen tetrad 55-65 μm diameter, spines and spinules well defined, spines 2 μm long, 7-12 μm²; ovary with 3 styles, bipartite near base. Seeds papillose, ellipsoid, papillae arranged irregularly, 0.50-0.55 mm x 0.25-0.30 mm.

Distribution: Common in Eurasia, and the coastal plain from southern Virginia to Texas (North America), Central America, Caribbean islands and northern South America: Venezuela, Guyana, Suriname and Brazil, expected in French Guiana; wet savanna, 400-1300 m alt.; 30 collections studied, 12 from the Guianas (GU: 7; SU: 5).

Selected specimens: Guyana: de la Cruz 4012 (US); Jenman 2255 (K). Suriname: Moengo tapoe, Grote Zwiebelzwamp, Lanjouw & Lindeman 853, 903, 944 (U); Tafelberg, Maguire 24432, 24486 (VEN, NY, U).

6. **Drosera kaieteurensis** Brumm.-Ding., Acta Bot. Neerl. 4: 137. 1955. Type: Guyana, Kaieteur Savanna, Maguire & Fanshawe 23466 (holotype NY, isotype U). – Fig. 34

Acaulescent herb forming rosette, no more than 2 cm tall. Leaves spathulate, 5.5-15 mm long, petiole and blade well defined; stipules 4-5 mm long, 3-4-partite; petiole 3-10 x 0.3-0.8 mm, hairy beneath; blade very widely obovate or widely obovate, 2.5-5 x 2.5-4 mm, hairy beneath. Inflorescences 1-5 per plant, 1-7 cm long, 1-5-flowered; peduncle erect, 1-4 cm long, with long non-glandular hairs and smaller glandular hairs. Flowers with sepals narrowly ovate, joined at base, 2-3 x 1-1.5 mm, reflexed at anthesis, margin entire, with long non-glandular hairs and smaller glandular hairs outside; petals white or pink, spathulate, 3 x 1.5 mm; stamens 2.5-3 mm long; pollen tetrad 50-65 μm diameter, spines very dense, 1.5 μm long, more than 50 μm^2; ovary with 3 styles, bipartite near base. Fruit after ripening cup-like; seeds ovate or subglobose, foveolate, 0.30-0.4 x 0.15-0.20 mm, covered with granular wax.

Distribution: Venezuela (Bolívar: Chimantá, Guaquinima, Auyan-tepui and Amuray-tepui) and Guyana (Pakaraima region); wet sandy savanna, bog, scrub vegetation and sandy areas by streams, 400-1900 m alt.; 21 collections studied, 9 from Guyana (GU: 9).

Selected specimens: Guyana: Pakaraima Mts, Mt. Membaru, Maas et al. 4268, 4346 (K, U); Kaieteur Plateau, Maguire & Fanshawe 23466 (NY, U); Imbaimadai, Membaru-Kurupung, Kamarang, Maguire & Fanshawe 32286 (K, US, VEN).

152

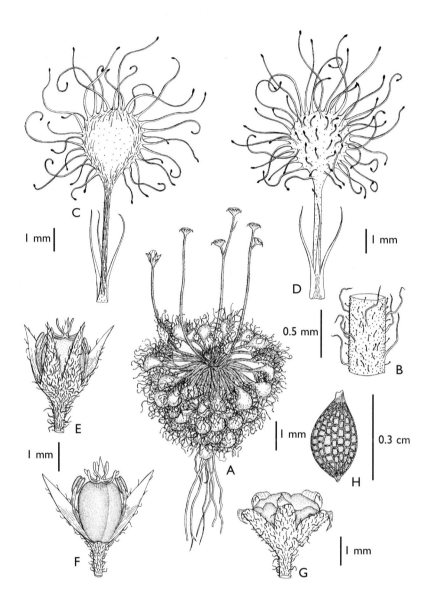

Fig. 34. *Drosera kaieteurensis* Brumm.-Ding.: A, habit; B, detail of peduncle; C, leaf, below; D, leaf, above; E, flower; F, opened flower; G, fruit after ripening; H, seed. (A-H, from Duno *et al.* 456). Drawing by Juan Castillo.

7. **Drosera roraimae** (Diels) Maguire & J.R. Laundon in Maguire & Wurdack, Mem. New York Bot. Gard. 9: 333. 1957. – *Drosera montana* A. St.-Hil. var. *roraimae* Diels in Engl., Pflanzenr. IV. 112 (Heft 26): 90. 1906. Type: Venezuela, Bolivar, Mt. Roraima, Ri. Schomburgk 1034 (holotype B, not seen). – Fig. 35

Drosera montana A. St.-Hil. var. *robusta* Diels, Notizbl. Bot. Gart. Berlin-Dahlem 6: 136. 1914. Type: Venezuela, Bolivar, Roraima, Ule 8610 (holotype, B, not seen).

Caulescent herb forming an apical rosette. Stem up to 10 cm long, invested by marcescent leaves. Leaves spathulate, 10-25 mm long, petiole and blade well defined; stipules coriaceous and conspicuous, 3-7 mm long, 3-7-partite; petiole 6-15 x 1-2 mm, conspicuously hairy beneath; blade obovate, 4-10 x 2-4 mm, hairy beneath. Inflorescences 1-2(-4) per plant, 8-30 cm long, (4-)6-20-flowered; peduncle erect or curved near base, 8-20 cm long, with glandular hairs or sessile glands, rarely glabrous. Flowers with sepals ovate or narrowly ovate, joined at base, 4-5 x 1-2 mm, margin entire, with glandular hairs outside; petals white or pink, spathulate, 4-6 x 2-2.5 mm; stamens 4-6 mm long; pollen tetrad 55-75 μm diameter, spines and spinules present, spine 2 μm long, 9-14 μm^2; ovary with 3 styles, bipartite near base. Seeds foveolate-reticulate, narrowly oblongoid, 0.4-0.6 x 0.15-0.30 mm, covered with granulate wax.

Distribution: Venezuela (Amazonas and Bolívar), Guyana (Pakaraima) and Brazil (Cerro de La Neblina); wet sand savanna, scrub vegetation and sandy areas by streams, 900-2700 m alt.; most common species in the Guayana Highland of Venezuela; more than 100 collections studied, 12 from Guyana (GU: 12).

Selected specimens: Guyana: Pakaraima Mts., Mt. Aymato, Maas *et al.* 5660 (NY, U); Mt. Ayanganna, Kaieteur Plateau, Maguire *et al.* 40645 (NY).

8. **Drosera sessilifolia** A. St.-Hil., Hist. Pl. Remarq. Brésil 1: 259, t. 25A. 1825-1826. Type: Brazil, Minas Gerais, near the San Franscisco R., Saint Hilaire B1 N 1805 (holotype P).

Drosera dentata Benth., J. Bot. (Hooker) 4: 105. 1841. Type: Guyana: Ro. Schomburgk ser. I, 102 (holotype K, not seen, isotype U).

154

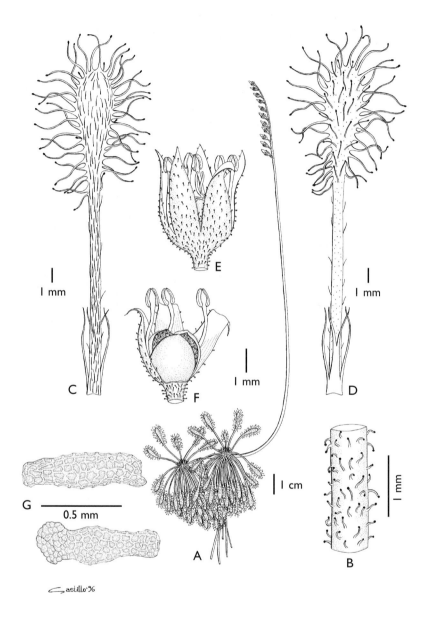

Fig. 35. *Drosera roraimae* (Diels) Maguire & J.R. Laundon: A, habit; B, detail of peduncle; C, leaf, below; D, leaf above; E, flower; F, fruit after ripening; G, seed. (A-G, from Trujillo 13679). Drawing by Juan Castillo.

Acaulescent herb forming a rosette, no more than 2 cm tall. Leaves oblanceolate (cuneate), 10-30 mm long, petiole grading into blade, indistinct; stipules 2-4 mm long, 5-7-partite. Inflorescences (1-)2-3(-5) per plant, 5-25 cm long, (3-)5-10(-15)-flowered; peduncle erect, 6-25 cm long, glabrous or with few glandular hairs. Flowers with sepals ovate or narrowly ovate, joined at base, 4-6 x 2 mm, margin entire or with glandular hairs, with glandular hairs outside; petals pink, obovate or spathulate, 5 mm long; stamens 4-4.5 mm long; pollen tetrad 55-65 μm diameter, spines and spinules well defined, spine 1.5 μm long, 15.5-20.5 μm^2; ovary with 5 simple styles. Seeds reticulate, subglobose, 0.3-0.5 mm long, wax not evident.

Distribution: Venezuela, Guyana, Suriname and Brazil, expected in French Guiana; common in wet sand savanna, 50-150 m alt.; most common species in lowland savanna in Venezuela; more than 50 collections studied, 13 from the Guianas (GU: 12; SU: 1).

Selected specimens: Guyana: Rupununi Savanna, Mountain Point, near ranch of Shirley Humphries, Jansen-Jacobs *et al.* 455 (U); Rupununi Savanna, near Mountain Point, just S. of Kanuku Mts., Maas & Westra 4022 (U). Suriname: Sipaliwini Savanna, van Donselaar 3664 (U).

WOOD AND TIMBER
by
Jifke Koek-Noorman[14] and Pierre Détienne[15]

PHYTOLACCACEAE

WOOD ANATOMY

In the Guianas, three genera occur that are known to include woody species. The only species represented in the Utrecht wood collection is the scandent shrub *Seguieria americana*.
As far as known the wood is of no commercial value.

FAMILY DESCRIPTION

Included phloem present in tangential concentric zones, consisting of bundles of xylem and phloem, separated by conjunctive parenchyma (Fig. 39 A); bands sometimes anastomosing, 200-400 μm wide, 5-6 per cm. Many cells filled with rhombic crystals or long styloids.
Vessels round to oval, solitary and in small multiples, diffuse, regularly distributed or lacking in narrow tangential zones at the outer side of the phloem bands; diameter 40-125 μm, small and large vessels intermingled; 25-40 vessels per mm². Perforations simple. Intervascular pits oval, 4 x 6 μm. Vessel-ray pits similar to intervessel pitting. Vessel member length 300-400 μm.
Rays 1-5-seriate, the multiseriates with short to long uniseriate margins; multiseriate parts up to 750 (2000) μm high, often slightly irregularly built, mainly composed of procumbent cells, the larger rays with sheath cells; margins like uniseriate rays composed of procumbent, square and upright cells; up to 1250 μm, 8 cells high, 5-8 per mm.
Parenchyma very scanty paratracheal as fusiform cells or 2-celled strands; strands 120-150 μm high; as conjunctive tissue in the alternating with the islands of included phloem.

[14] Nationaal Herbarium Nederland, Utrecht University branch, Heidelberglaan 2, 3584 CS Utrecht, The Netherlands.
[15] C.I.R.A.D.-Forêt, TA 10/16, rue Jean François Breton, 34398 Montpellier Cedex 5, France.

Acknowledgements: Technical assistance was given by D. Makhan and H. Rypkema.
Figs. 36 A, B and C: courtesy of L.Y.Th. Westra.

Fibres with moderately thick-walled, walls 3-4 μm, lumina up to 15 μm wide. Pits simple, slit-like, numerous on radial walls, scanty on tangential walls.

Crystals of different size and shape, varying from rhombic and fusiform crystals to styloids in the included phloem (Carlquist 1988).

Material studied (Uw-numbers refer to the Utrecht Wood collection): **Seguieria americana** L.: Suriname: Heyde 626 (Uw 23242; diam. 2 cm); Heyde & Lindeman 124 (Uw 22759; diam. 7.5 cm).

NYCTAGINACEAE

WOOD ANATOMY

FAMILY CHARACTERISTICS

The family can easily be recognised by some anatomical features, in particular abnormal secondary growth by the included phloem, is easily spotted with a hand lens (Fig. 36 A, B). See also Metcalfe & Chalk 1983, p. 211; Carlquist 1988, p. 256; Lindeman *et al.* 1963, p. 249). In the wood of *Guapira*, *Neea* and *Pisonia*, phloem is found as islands that abaxially accompany short, irregular radial rows or chains of small to medium-sized vessels. These conspicuous structures, formed by vessels, parenchyma strands and abaxial phloem islands are diffusely spread throughout the wood. The rays, consisting of procumbent to square cells, are narrow, uniseriate with accidentally low, biseriate parts.

In the wood of *Guapira*, *Neea* and *Pisonia*, the nyctaginaceous genera indigenous in the Guianas, some variation between the studied species was found in the size and number of woodanatomical characters, i.e. the size of the phloem islands, the vessel diameter, the height of the rays, and the amount of fibre pits (see remarks in the family description). It is, however, not possible to distinguish between these genera.

The family description is based on wood samples of *Guapira*, *Neea* and *Pisonia*.

Although *Bougainvillea* is not indigenous for the Guianas, it is very often found as ornamental plant. *Bougainvillea* is deviating in several aspects. The phloem islands, typical for the other Nyctaginaceae, are here also found. They are, however, arranged in tangential zones, at the adaxial side of continuous, 4-6 cells wide parenchyma bands (Fig. 36 B, 38 C). Towards the bark, each parenchyma bands is bounded by a zone without vessels.

158

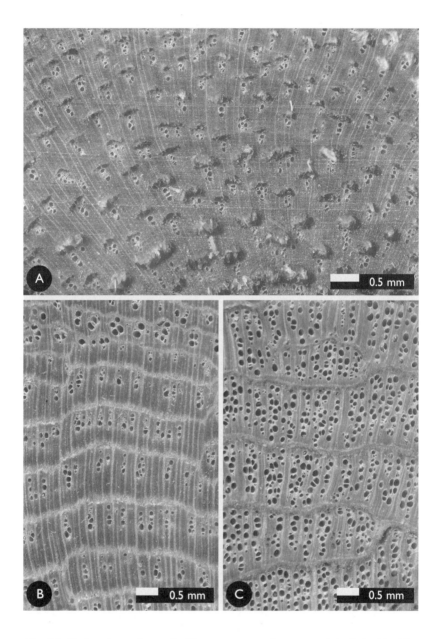

Fig. 36. End-grain surfaces. A, *Neea constricta* Spruce ex J.A. Schmidt (Uw 33504); B, *Bougainvillea glabra* Choisy (Uw 12632); C, *Seguieria americana* L. (Uw 22759).

In *Bougainvillea* the rays are high, slightly irregular, mostly up to 4 cells wide, often with long uniseriate margins, consisting of upright and some rows of square cells. Many ray cells are of appr. the same height as the vessel members and parenchyma cells and contribute to the ripple marks (Fig. 38 D).

FAMILY DESCRIPTION

Included phloem present as islands of 500/600 x 375/550 μm, islands oval (to round), rarely kidney-shaped, diffusely spread (Fig. 37 A, 38 A). Often raphids or styloids included (Fig. 37 C).

Vessels at the adaxial side of the phloem islands, as irregular or regular radial rows and chains up to 8 vessels long, intermingled with few parenchyma strands (Fig. 37 A); 1-3(-5) vessel groups per mm^2 (5-10 vessels per mm^2); vessels round to oval, diameter 50-100(-150) μm. Perforations simple. Intervascular pits alternate, round to slightly oval, 4-6 μm. Vessel-ray pits similar to intervascular pits or slightly larger. Vessel member length ca. 150-200 μm.

Rays uniseriate up to 14-18 cells, 350-500 μm high, consisting of procumbent to nearly square cells, with scanty biseriate parts of 1-4 cells high (Fig. 37 B), 7-12 per mm.

Parenchyma as fusiform cells and strands of 2 cells (Fig. 37 B), ca. 150 μm high; paratracheal as scanty cells in the vessel groups, apotracheal as one cell wide layers, abaxially accompanying the phloem islands.

Fibres thin- to thick-walled, walls ca. 3-6(-8) μm, lumen (6-)18-20 μm wide. Pits simple, slit-like, the occurrence on both radial and tangential walls varying from scanty to numerous.

Ripple marks present to absent in *Guapira*, formed by fibres, parenchyma and vessels members, rarely also by rays (Fig. 37 B); very vague to absent in *Neea* and *Pisonia*.

R e m a r k s : The vessel diameter is up to 50 μm in *Neea mollis* (Fig. 38 A) and in the Guianan material of *N. spruceana*. The vessel diameter of the Krukoff collections of *N. spruceana* from Brazil however, is (50-)85-90(-115) μm.

In some species (i.e. *Guapira eggersiana*) only square ray cells were observed. The rays of *Pisonia macranthocarpa* are deviating by their height: (1000-1200 μm; up to 40 cells) with biseriate parts of up to 6 cells; these rays are terminally or laterally accompanied by upright cells (Fig. 38 B).

Short irregular, narrow apotracheal bands of parenchyma intermingled with thinwalled fibres are present in *Guapira eggersiana*. Diffuse

160

Fig. 37. A, *Guapira eggersiana* (Heimerl) Lundell: transverse section (Uw 21276); B, *Guapira eggersiana* (Heimerl) Lundell: tangential section (Uw 21276); C, *Guapira cuspidata* (Heimerl) Lundell: tangential section (Uw 1385).

apotracheal parenchyma is abundant in *Pisonia macranthocarpa* in short, one cell wide bands; fusiform cells are also found as ray margins or "sheath cells".

TIMBERS AND THEIR PROPERTIES

Botanical	The woods of the two genera *Guapira* and *Neea* are similar and treated together.
Tree	Small to medium-sized trees. Only *Guapira kanukuensis* reported as large tree.
Description of the wood	No distinction between sapwood and heartwood, pale yellow, whitish to grey. Texture very coarse, the included phloem islands seeming as large vessels on longitudinal sections. Grain straight. Luster low.
Weight	Generally 600-800 kg/m³ (*G. eggersiana, N. floribunda* and most of unidentified nyctagineous wood samples), but 480-600 kg/m³ in *G. cuspidata* and *N. spruceana,* or up to 1000 kg/m³ in *G. salicifolia* and *N. ovalifolia.*
Uses	Any use reported and no commercial possibilities (Record & Hess 1943, p. 410)

Material studied: (CTFw-numbers refer to the collection of CIRAD-Forêt Montpellier, Uw-numbers to the Utrecht Wood collection): **Bougainvillea glabra** Choisy: Brazil: Lindeman & de Haas 550 (Uw 12632), 750 (Uw 12725).
Guapira cuspidata (Heimerl) Lundell: Guyana: Forest Dept. British Guiana 5387 = F 2599 (Uw 1013). Suriname: Lanjouw & Lindeman 1084 (Uw 1385), Lindeman, Görts *et al.* 458 (Uw 26446).
G. eggersiana (Heimerl) Lundell: Suriname: Stahel 29 (Uw 29), Lanjouw & Lindeman 1082 (Uw 1383), Lindeman '53-'55 5416 (Uw 3761), Mennega *et al.* 887 (Uw 21275).
G. kanukuensis (Standl.) Lundell: Guyana: A.C. Smith 3594 (Uw 21702). Suriname: BBS 102 (Uw 696).

162

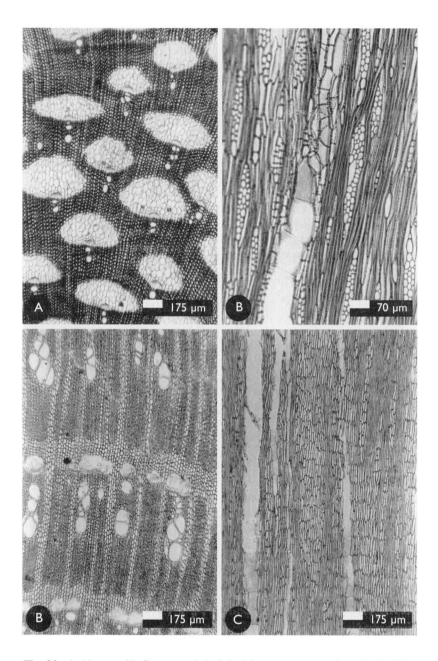

Fig. 38. A, *Neea mollis* Spruce ex J.A. Schmidt: transverse section (Uw 21219); B, *Pisonia macranthocarpa* (Donn. Sm.) Donn. Sm.: tangential section (Uw 20744); C, *Bougainvillea glabra* Choisy: transverse section (Uw 12632); D, *Bougainvillea glabra* Choisy: tangential section (Uw 12632).

G. salicifolia (Heimerl) Lundell: Guyana: Jansen-Jacobs *et al.* 3676 (Uw 34811). Suriname: Maas & Tawjoeran LBB 10943 (Uw 11686).
Neea constricta Spruce ex J.A. Schmidt: Guyana: Stoffers, Görts-van Rijn *et al.* 154 (Uw 30099). Suriname: Lanjouw & Lindeman 2549 (Uw 1821). French Guiana: Feuillet *et al.* 10218 (Uw 33504).
N. floribunda Poepp. & Endl.: Guyana: Stoffers, Görts-van Rijn *et al.* 305 (Uw 30146). Suriname: Lanjouw & Lindeman 1378 (Uw 1468), 1463 (Uw 1494). French Guiana: Sabatier *et al.* 2150 (CTFw 31287, Uw 32772).
N. mollis Spruce ex J.A. Schmidt: French Guiana: Maas *et al.* 2238 (Uw 21219).
N. ovalifolia Spruce ex J.A. Schmidt: Guyana: A.C. Smith 2913 (Uw 21596). Suriname: Mennega 495 (Uw 2983), Lindeman 4900 (Uw 3339). Brazil: Krukoff 7958 (CTFw 8386).
N. spruceana Heimerl: Suriname: Lanjouw & Lindeman 2828 (Uw 1943a). French Guiana: de Granville *et al.* 7955 (Uw 31527). Brazil: Krukoff 8102 (CTFw 8385). Peru: Williams 623 (CTFw 8230).
Pisonia macranthocarpa (Donn. Sm.) Donn. Sm.: ?: Madison 20989 B7-A (Uw 20744).

AMARANTHACEAE

WOOD ANATOMY

The only species occurring in the Guianas and represented in the Utrecht wood collection is the climbing liana *Chamissoa altissima*. The xylem body is divided in small parts by a network, formed by unlignified ray-like zones and wide tangential bands of unlignified conjunctive tissue (Fig. 39 B).
As far as known the wood is of no commercial value.

FAMILY DESCRIPTION

Included phloem included in tangential bands of conjunctive tissue which form a reticulate pattern with the unlignified rays.
Vessels diffuse, solitary or in small multiples, round to oval, diameter 40-150 μm, 10-30 per mm^2, the variation mainly due to the unlignified vesselless zones.

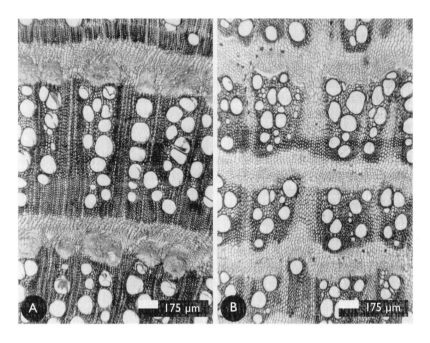

Fig. 39. A, *Seguieria americana* L.: transverse section (Uw 23242); B, *Chamissoa altissima* (Jacq.) Kunth: transverse section (Uw 32171).

Perforations simple. Intervascular pits alternate, round, 7-8 μm. Vessel-ray pits simple, oval to elongated, slightly irregular. Vessel member length 300-400 μm.

Ray-like zones of mainly unlignified parenchymatic cells linking the successive layers of conjunctive tissue.

Crystal sand present in radial and tangential conjunctive tissue.

Parenchyma scantily diffuse and as narrow vasicentric sheaths; fusiform cells and strands of 1-2 cells, 200-300 μm high.

Fibres thin-walled, walls 2-4 μm, lumina up to 25 μm wide. Pits minute, simple, numerous on both tangential and radial walls.

Material studied (Uw-numbers refer to the Utrecht Wood collection): **Chamissoa altissima** (Jacq.) Kunth: Guyana: Jansen-Jacobs *et al.* 960 (Uw 32171; diam. 1.5 cm), 1178 (Uw 32240; diam. 1.2 cm).

LITERATURE ON WOOD AND TIMBER

Carlquist, S. 1988. Comparative Wood Anatomy. Systematic, Ecological, and Evolutionary Aspects of Dicotyledon Wood. Springer-Verlag, Berlin-Heidelberg.

Lindeman, J.C., A.M.W. Mennega and W.J.A. Hekking. 1963. Bomenboek voor Suriname. Herkenning van Surinaamse houtsoorten aan hout en vegetatieve kenmerken. Uitgave Dienst 's Lands Bosbeheer Suriname - Paramaribo.

Metcalfe, C.R. & L. Chalk. 1983. Anatomy of the Dicotyledons vol. II. Wood structure and conclusion of the general introduction. Clarendon Press, Oxford.

Record, S.J. & R.W. Hess. 1943. Timbers of the New World. Yale University Press, New Haven, CT.

NOMENCLATURAL NOVELTIES

Nyctaginaceae

A lectotype is selected for *Pisonia glabra* (taxonomical synonym of Guapira eggersiana).

NUMERICAL LIST OF ACCEPTED TAXA

Phytolaccaceae

1. Microtea Sw.
   1-1. M. debilis Sw.
   1-2. M. maypurensis (Kunth) G. Don

2. Petiveria L.
   2-1. P. alliacea L.

3. Phytolacca L.
   3-1. P. rivinoides Kunth & C.D. Bouché
   3-2. P. thyrsiflora Fenzl ex J.A. Schmidt

4. Rivina L.
   4-1. R. humilis L.

5. Seguieria Loefl.
   5-1. S. americana L.
   5-2. S. macrophylla Benth.

6. Trichostigma A. Rich.
   6-1. T. octandrum (L.) H. Walt.

Nyctaginaceae

1. Boerhavia L.
   1-1. B. diffusa L.

2. Bougainvillea Comm. ex Juss.
   2-1. B. glabra Choisy
   2-2. B. spectabilis Willd.

3. Guapira Aubl.
   3-1. G. cuspidata (Heimerl) Lundell
   3-2. G. eggersiana (Heimerl) Lundell
   3-3. G. kanukuensis (Standl.) Lundell
   3-4. G. salicifolia (Heimerl) Lundell

4. Mirabilis L.
   4-1. M. jalapa L.

5. Neea Ruiz & Pav.
   5-1. N. constricta Spruce ex J.A. Schmidt
   5-2. N. floribunda Poepp. & Endl.
   5-3. N. mollis Spruce ex J.A. Schmidt
   5-4. N. ovalifolia Spruce ex J.A. Schmidt
   5-5. N. spruceana Heimerl

6. Pisonia L.
   6-1. P. macranthocarpa (Donn. Sm.) Donn. Sm.

Aizoaceae

1. Sesuvium L.
   1-1. S. portulacastrum (L.) L.

Chenopodiaceae

1. Chenopodium L.
   1-1. C. ambrosioides L.

Amaranthaceae

1. Achyranthes L.
   1-1. A. aspera L. var. aspera

2. Alternanthera Forssk.
   2-1. A. brasiliana (L.) Kuntze
   2-2. A. ficoidea (L.) P. Beauv.
   2-3. A. halimifolia (Lam.) Standl. ex Pittier
   2-4. A. philoxeroides (Mart.) Griseb.
   2-5. A. sessilis (L.) DC.

3. Amaranthus L.
   3-1. A. australis (A. Gray) J.D. Sauer
   3-2. A. blitum L.
   3-3. A. caudatus L.
   3-4. A. dubius Mart. ex Thell.
   3-5. A. hybridus L.
   3-6. A. spinosus L.
   3-7. A. viridis L.

4. Blutaparon Raf.
   4-1. B. vermiculare (L.) Mears var. vermiculare

5. Celosia L.
   5-1. C. argentea L. f. argentea

6. Chamissoa Kunth
   6-1. C. acuminata Mart. var. swansonii Sohmer
   6-2. C. altissima (Jacq.) Kunth var. altissima

7. Cyathula Blume
   7-1. C. achyranthoides (Kunth) Moq.
   7-2. C. prostrata (L.) Blume

8. Gomphrena L.
   8-1. G. globosa L.

9. Iresine P. Browne
   9-1. I. diffusa Humb. & Bonpl. ex Willd.

10. Pfaffia Mart.
    10-1. P. glabrata Mart. var. rostrata O. Stützer
    10-2. P. glomerata (Spreng.) Pedersen var. glomerata
    10-3. P. grandiflora (Hook.) R.E. Fr. var. grandiflora

Portulacaceae

1. Portulaca L.
   1-1. P. grandiflora Hook.
   1-2. P. oleracea L.
   1-3. P. pilosa L.
   1-4. P. quadrifida L.
   1-5. P. sedifolia N.E. Br.

2. Talinum Adans.
   2-1. T. fruticosum (L.) Juss.
   2-2. T. paniculatum (Jacq.) Gaertn.

Basellaceae

1. Anredera Juss.
   1-1. A. leptostachys (Moq.) Steenis

2. Basella L.
   2-1. B. alba L.

Molluginaceae

1. Glinus L.
    1-1. G. radiatus (Ruiz & Pav.) Rohrb.

2. Mollugo L.
    2-1. M. verticillata L.

Caryophyllaceae

1. Drymaria Schult.
    1-1. D. cordata (L.) Schult.

2. Polycarpaea Lam.
    2-1. P. corymbosa (L.) Lam.

Sarraceniaceae

1. Heliamphora Benth.
    1-1. H. nutans Benth.

Droseraceae

1. Drosera L.
    1-1. D. biflora Humb. & Bonpl. ex Roem. & Schult.
    1-2. D. capillaris Poir.
    1-3. D. cayennensis Sagot ex Diels
    1-4. D. esmeraldae (Steyerm.) Maguire & Wurdack
    1-5. D. intermedia Hayne
    1-6. D. kaieteurensis Brumm.-Ding.
    1-7. D. roraimae (Diels) Maguire & J. R. Laundon
    1-8. D. sessilifolia A. St.-Hil.

COLLECTIONS STUDIED
(Numbers in bold represent types)

Phytolaccaceae

GUYANA

Appun, C.F., 327 (1-1); 1787 (5-1); 2459 (1-2)
Archer, W.A., 2256 (1-1); 2295 (3-1)
Botanical Garden, s.n. (2-1)
Clarke, H.D., 978 (3-1)
Cooper, A., 32 (2-2)
Cruz, J.S. de la, 1143 (1-1); 2085 (3-1); 2093 (1-1); 2132 (3-1); 2480, 2501 (1-1); 3130, 3221 (3-1); 3548 (1-1); 3639 (3-1); 3650, 4033 (1-1); 4481 (3-1); 4601 (1-1)
Davis, D.H., 893 (5-1)
Fairchild, D., s.n. (6-1)
Fanshawe, D.B., 2450=FD 5186 (3-1)
FD (Forest Dept. British Guiana, see also Fanshawe), 5186=F 2450 (3-1)
Gillespie, L.J. et al., 1027, 1313 (3-1)
Gleason, H.A., 18 (1-1); 55, 194 (3-1); 696 (1-1); 826 (3-1)
Gorinsky, C., 21 (2-1)
Grewal, M.S. & R. Persaud, 94 (3-1)
Hahn, W. et al., 3903 (2-2); 5676 (3-1)
Harrison, S.G. & R. Persaud, 1072 (1-1)
Henkel, T.W., 630 (1-1); 4150, 5988 (3-1)
Hitchcock, A.S., 16587 (4-1); 16608 (2-1); 17115 (1-1); 17383 (3-1)

Hoffman, B., 2419 (3-1)
Irwin, H.S., 237 (1-1); 788 (1-2); **797** (5-1)
Jansen-Jacobs, M.J. et al., 236 (5-1); 1005, 1546 (3-1); 3835 (4-2); 4250 (5-1); 4733 (1-1)
Jenman, G.S., 503 (4-2); 2013 (3-1); 2192, 5151 (6-1); 5277 (1-1); 6524 (3-1); 7849 (1-1)
Kvist, L.P. et al., 5, 259 (3-1)
Lee, G. & R. Persaud, 18 (2-2)
McDowell, T. et al., 2168 (2-1); 2627 (3-1)
Mell, C.D. & R.C. Mell, 234 (1-1)
Mutchnick, P., 1501 (3-1)
Omawale & R.N. Persaud, 217 (3-1)
Pipoly, J.J. et al., 7924 (2-2); 8307 (3-1); 9716 (2-2)
Ramsammy, J.R., 9 (3-1)
Roberston, K.R. & D.F. Austin, 282 (2-2); 291 (4-1)
Sandwith, N.Y., 1566 (3-1)
Schomburgk, Ri., 47 (1-1); 203 (3-1)
Schomburgk, Ro., s.n. (1-1); s.n. (1-2); s.n. (2-1); ser. I, 116 (1-2); ser. I, 229 (1-1); 470 (K) (2-2); **ser. I, 661** (5-1); ser. I, 835 (1-2); ser. I, 860 (2-1)
Smith, A.C., 3650 (4-2)
Steege, H. ter et al., 224 (2-2); 439 (3-1)
Stockdale, F.A., s.n. (1-2)
Stoffers, A.L. & A.R.A. Görts-van Rijn et al., 193 (3-1)
Tate, G.H.H., 299 (2-2)
Warren, L.B., s.n. (1-1)

Fleury, M., 451 (2-1); 492 (1-1)
Fournet, A., 139 (1-1); 198 (3-1)
Garnier, F.A., 102 (3-1)
Gely, A., 43 (2-1)
Granville, J. J. de, *et al.*, 1150 (3-1); 2532 (2-1); 3649, 7826, 8703, 10722 (3-1)
Grenand, P., 53 (2-1); 201 (3-1); 1117 (5-1); 1670 (3-1); 2875 (2-1)
Hoff, M., 6083 (1-1); 6252 (2-2)
Jacquemin, H., 1467 (3-1); 1498 (1-1); 2842 (2-1)
Kodjoed, J.F., 92 (2-1); 99 (1-1)
Larpin, D., 453 (3-1)
Leclerc, A., 0051 (1-1)
Lescure, J.P., 310 (3-1)
Moretti, C., 119, 585, 734, 1294 (2-1)
Mori, S. A. *et al.*, 18246 (2-1); 18639, 19065, 21186 (3-1); 23295, 23818 (1-1)
Oldeman, R.A.A., B-438 (3-1); 1259 (2-2); 1651 (3-1); 3120, 3151 (5-1); B-3892 (2-1); B-4019 (3-1)
Poiteau, P.A., s.n. (3-1)
Prévost, M.F. *et al.*, 717 (3-1); 746 (1-1); 892 (5-1); 1356 (1-1)
Riera, B., 930 (3-1)
Sastre, C. & F. Sastre, 282 (2-2); 4041 (3-1)
Sauvain, M., 768 (2-1)
Service Forestier, s.n. (1-1); **6028** (2-2)
Skog, L. & C. Feuillet, 7300 (3-1)
Taverne, B., 88 (2-1)
Weitzman, A. & W. Hahn, 318 (3-1)
Wittingthon, V., 58 (2-1)
without coll., 1865 (3-1)

Nyctaginaceae

GUYANA

Acevedo, P., 3474 (3-2)
Anderson, C.W., 393 (3-1); 612 (5-2)
Bartlett, A.W., s.n., 241, 8597 (3-4)
Boyan, J., 30=FD 7714 (3-2)
Cruz, J.S. de la, 965 (4-1); 3017, 3315 (3-2); **3317** (3-4); 4120 (3-2); 4557 (3-4)
Davis, D.H., 869 (1-1); 1669 (3-4)
Ek, R.C. *et al.*, 633 (5-2); 674 (5-1)
Fanshawe, D.B., s.n. (3-4); 291, 310 (5-2); 329=FD 3065 (3-2); 331 (5-4); 511 (5-2); 1137=FD 3873 (5-3); 1435=FD 4171 (5-1); 1463=FD 4199, 1747 (5-2); 2140=FD 4876 (5-1); 2338=FD 5074, 2347 (3-2); 2378 (3-4); 2383 (3-2); 2599=FD 5387 (3-1); 2677 (3-4); 2697=FD 5490 (5-3); 2775 (5-1); 2956=FD 6286 (3-1); 3027, 3046, 3247 (5-2); 3337 (3-2); 3364, 3366, 3368, 3371, 3393 (5-4); 3395, 3399 (3-1); 4483 (5-2); 5574 (5-1)
Forest Dept. British Guiana (FD), (see also Boyan, Fanshawe, Guppy, Persaud, Pinkus and Wilson-Browne), 2815=P 43, 3065=F 329 (3-2); 3873=F 1137 (5-3); 4171=F 1435 (5-1); 4199=F 1463 (5-2); 4876=F 2140 (5-1); 5074=F 2338 (3-2); 5387=F 2599 (3-1); 5490=F 2697 (5-3); 5944=WB 545 (3-2); 6286=F 2956 (3-1); 6779 (1-1); 7675=G 660 (5-4); 7714=JB 30 (3-2)

SURINAME

Hostmann, F.W.R. *et al.*, 390 (1-1); 414 (3-2); 759a (1-1); **824**, **947** (3-2)

Hulk, J.F., 84 (4-1)

Jansma, R., 23 (3-2)

Jonker-Verhoef, A.M.E. & F.P. Jonker, 513 (3-4)

Julen, C.R., in LBB 14552 (3-2); 14572 (1-1)

Kanhai, E.D., in LBB 13346 (3-2)

Kramer, K.U. & W.H.A. Hekking, 2067, 2158 (3-2); 3098 (1-1)

Kuyper, J., 36 (1-1)

Lanjouw, J. & J.C. Lindeman, 637 (1-1); 1042 (3-2); 1047 (1-1); 1082 (3-2); 1084 (3-1); 1123 (3-2); 1134 (3-4); 1142, 1203, 1366 (3-2); 1378 (5-2); 1380 (3-4); 1382 (3-2); 1440 (1-1); 1463 (5-2); 1781 (3-2); 2402 (5-2); 2586 (5-2); 3117 (3-2)

LBB (Lands Bosbeheer, see also Boerboom, Elburg, Julen, Kanhai, Maas, Narain, Pons, Sabajo, Schulz, Sterringa, Teunissen and Tjon Lim Sang), 9406 (5-2); 10533 (1-1); 12131 (5-4); 12144 (3-2)

Lindeman, J.C., 3851, 3874, 3917 (3-2); 4900, 4978 (5-4); 5102, 5302, 5416 (3-2); 5649 (5-3); 5670 (3-2); 5685 (1-1); 5807, 5826 (5-2); 6641 (3-2); 6884 (3-2)

Lindeman, J.C., A.R.A. Görts-van Rijn *et al.*, 458 (3-1); 623 (5-4)

Maas, P.J.M. & J. A. Tawjoeran, in LBB 10943 (3-4)

Maguire, B. & G. Stahel, 22741 (1-1); 22793, 24628, 24955 (3-2)

Mennega, A.M.W. *et al.*, 495 (5-4); 887 (3-2)

Mori, S.A. & A.B. Bolten, 8481, 8497, 8577 (5-4)

Narain, T.R., in LBB 13761 (1-1)

Outer, R.W. den, 955 (3-2)

Pons, T.L., in LBB 12653 (1-1)

Reijenga, T.W., 422 (1-1)

Sabajo, P.H., in LBB 9061, 11201 (5-4)

Samuels, J.A., s.n. 122, (1-1)

Schomburgk, Ri., s.n. (3-2)

Schulz, J.P., 7852 (5-2); 7875 (5-4); 8020 (5-2);

Schulz, J.P. *et al.*, in LBB 9909, 9952 (3-2)

Soeprato, D5, B9 (3-2); B19 (1-1); E47 (3-2)

Stahel, G. (Woodherb. Sur.), 29a, 55 (3-2); 96 (5-2); 253 (5-4); 253a (3-4)

Sterringa, J.T., in LBB 12417, 12518 (3-2)

Teunissen, P.A., in LBB 15062, 15133, 15167, 15168, 15988, 16076 (3-2)

Tjon Lim Sang, R.J.M. & I.H.M. van de Wiel, 101b (3-2)

Tjon Lim Sang, R.J.M., in LBB 16242 (5-2)

Tresling, J.H.A.T., 151 (1-1)

UVS (Universiteit Suriname, see Werkhoven)

Went, F.A.F.C., 21 (4-1); 180 (1-1)

Werkhoven, M. *et al.*, in UVS 16457 (5-2); 16471 (5-3)

Wessels Boer, J.G., 535 (1-1)

without coll., 759 (1-1)

FRENCH GUIANA

Allorge, L., 351 (5-3)

BAFOG (Bureau Agricole et Forestier de Guyane, see Béna)

Barrier, S., 4038 (3-2); 4099 (5-2)

Béna, P., in BAFOG 1061 (3-2)

Benoist, R., s.n. (1-1); 556, 1299 (3-2); 1490 (1-1)

Billiet, F.P. & B. Jadin, 4756, 4758 (3-2

Boom, B.M. & S.A. Mori, 1941, 1942, 1946, 2091, 2176, 2268 (5-2); 2308 (5-4)

Bordenave, B., 338 (5-2)

Broadway, W.E., 353 (3-2); 698, 829 (3-2)

Cremers, G.A., *et al.*, 6669, 7004 (5-4); 9702 (3-2); 10494, 11417 (1-1); 11557 (5-2); 12906 (3-2)

Feuillet, C. *et al.*, 1198 (3-2); 10033 (5-3); 10078 (5-2); 10188 (5-1); 10218 (5-4)

Foresta, H. de, 681 (5-4); 686 (5-3)

Garnier, -, 110 (5-3)

Granville, J.J. de *et al.*, 456 (3-2); 652 (5-5); 837 (5-3); 1027 (5-4); T. 1163 (5-3); 2666 (3-2); 2769, 3047 (3-4); 3147, 3148 (5-3); 3326 (3-2); 3533, 3568 (5-3); 3589 (5-2); 3606 (5-3); 3650 (5-2); 3970 (5-4); 4252 (3-2); 4276 (5-2); 4843, 4911 (5-4); B. 5091 (5-4); 5115 (3-2); 5128, B. 5162 (5-3); 5764, 7098 (3-2); 7218 (5-4); 7260 (3-2); 7955 (5-5); 8228 (3-2); 8447, 11961 (5-4)

Grenand, P., 232 (5-4); 287, 468 (5-3); 1024 (5-4); 1744, 2099 (5-2); 2107 (3-4)

Hallé, F., 1023 (3-4)

Hoff, M., 5072 (2-2)

Jacquemin, H., 1442 (5-3); 2293 (5-2)

Jelski, C. von, **3100** (3-2)

Larpin, D., Nx 478 (3-2); Nx 522, 524, 671 (3-2); 965 (5-4)

Leprieur, D., s.n. (3-2); 185 (1-1)

Maas, P.J.M. *et al.*, 2238 (5-3)

Martin, J., s.n., s.n. (3-2)

Mélinon, E., 279 (1-1); 291 (5-4)

Middletown, Capt., 185 (1-1)

Moretti, C., 341 (5-3); 530 (3-4); 1105 (3-2)

Mori, S.A. *et al.*, 8756 (5-2); 8980 (3-4);14938 (5-3); 15193, 15333 (5-2); 18626 (5-4); 18754 (5-3); 21198 (5-1); 21636 (5-3); 22686 (5-2); 22787 (3-4); 22886 (5-2); 23208 (5-3); 23217 (5-1); 23747 (5-3); 23949 (5-4); 24205 (5-1)

Oldeman, R.A.A., T-152, T-172 (5-4); T-266 (3-2); 1050 (3-2); 1292 (5-4); B-1786 (5-3); 1811 (5-2); B-1865 (5-4); B-1895 (5-2); B-2587 (5-4); 2780 (5-4); 2816 (5-4); B-2852 (5-1); B-4004 (5-4)

Poiteau, P.A., s.n. (3-2)

Poncy, O., 945 (5-2)

Prance, G.T. *et al.*, 30653 (5-3)

Prévost, M.F. *et al.*, 2196, 2249 (5-4); 2851 (5-3)

Riera, B. *et al.*, 1581 (5-4); 1936 (3-2)

Sabatier, D. *et al.*, 1133, 1279 (5-4); 1584 (5-2); 1877 (3-4); 2150 (5-2); 2760 (5-2); 4195 (3-2)

Sagot, P.A., s.n. (1-1); s.n., s.n., s.n. (3-2); 990, 1017 (3-2)

Sarthou, C., 454 (3-2)

Sastre, C., 6309, 6464 (5-2)

Sauvain, M., 4 (5-5)

Service Forestier Guyane Française, 7082, 7279 (5-2)

Skog, L.E. & C. Feuillet, 7231 (5-2)

Soubirou, G., s.n. (1-1)

Villiers, J.F., 3708 (5-2)

Wachenheim, H., 336 (3-2)

without coll., 220 (1-1)

Aizoaceae

GUYANA

Grewal, M.S. & H. Lall, 293 (1-1)
Harrison, S.G. & R. Persaud, 1718 (1-1)
Hitchcock, A.S., 16567 (1-1)
Irwin, H.S., 294 (1-1)
Jenman, G.S., 5635, 5847 (1-1)
Leechman, A., s.n., s.n. (1-1)
Mutchnick, P., 372 (1-1)
Pipoly, J.J., 11245 (1-1)

SURINAME

Florschütz, P.A. *et al.*, 1913 (1-1)
Focke, H.C., 272, 557 (1-1)
Jonker-Verhoef, A.M.E. & F.P. Jonker, 515, 660 (1-1)
Kramer, K.U., 2108 (1-1)
Lanjouw, J., 547, 1100 (1-1)
Lanjouw, J. & J.C. Lindeman, 1060, 1066, 1509, 1522 (1-1)
LBB (Lands Bosbeheer, see Pons, Schulz and Sterringa)
Pons, T.L., in LBB 12644 (1-1)
Pulle, A., II 373 (1-1)
Reijenga, Th.W., 8, 8a, 9 (1-1)
Schulz, J.P., in LBB 10524 (1-1)
Soeprato, 80, 378 (1-1)
Sterringa, J.T., in LBB 12385 (1-1)
Wessels Boer, J.G., 530 (1-1)
without coll., s.n. (1-1)

FRENCH GUIANA

Cremers, G.A., 687, 8399, 8741, (1-1)

Descoings, B. & C. Luu, 20279 (1-1)
Goff, A. le & M. Hoff, 155 (1-1)
Granville, J.J. de, 1953, 8217 (1-1)
Hoff, M., 5496 (1-1)
Rossignol, -, 5 (1-1)
Rothery, H.C., 110 (1-1)
Skog, L. & C. Feuillet, 7541 (1-1)
Tostain, O.,73 (1-1)

Chenopodiaceae

GUYANA

Bartlett, A.W., 7932 (1-1)
Cook, C.D.K., 250 (1-1)
Davis, D.H., 911 (1-1)
Henkel, T.W. *et al.*, 3467 (1-1)
Irwin, H.S. *et al.*, 474 (1-1)
without coll., DP3M (1-1)

SURINAME

Focke, H.C., 1395 (1-1)

FRENCH GUIANA

Fleury, M., 860 (1-1)
Grenand, P., 69 (1-1)
Jacquemin, H., 1499, 1653, 2072 (1-1)
Kodjoed, J.F., 91 (1-1)
Moretti, C., 913 (1-1)
Oldeman, R.A.A., B-3909 (1-1)
Rothery, H.C., 59 (1-1)
Sauvain, M., 791 (1-1)
Taverne, B., 6 (1-1)
Wittingthon, V., 59 (1-1)

Amaranthaceae

GUYANA

Acevedo, P., *et al.*, 3462 (2-5); 3497 (6-2)

Bartlett, A.W., 8024 (2-4)

Boom, B.M., 7166 (4-1); 7174 (2-4)

Forest Bur. Suriname, BW G 224 (7-2); 423 (1-1)

Clarke, D., 382 (10-1); 1998 (2-1)

Cook, C.D.K., 252 (3-7)

Cowan, R.S., 39400 (6-2)

Cruz, J.S. de la, 964 (8-1); 973 (7-1); 982 (8-1); 1078 (2-5); 1245 (9-1); 1515 (2-3); 1672 (2-5); 1688 (2-2); 2126 (8-1); 2774 (9-1); 3579 (8-1); 3698 (7-1); 3751 (8-1); 3785, 3970 (9-1); 3997 (7-1); 4030 (2-5); 4078 (3-7)

Forest Dept. British Guiana, WB 445 (2-1)

Gleason, H.A., 280, 378, 788 (7-1)

Goodland, R., 822 (2-1)

Grewal, M.S. & H. Lall, 303 (4-1); 305 (3-1); 306 (2-5)

Hahn, W., 4840 (1-1)

Harris, S.A., M9 (7-2); TP 280 (3-4)

Harrison, S.G., 879 (1-1); 1334 (2-1)

Henkel, T.W. *et al.*, 622, 2944 (7-2)

Hiepko, P., 2775 (2-4)

Hitchcock, A.S., 16591 (3-2); 16609 (3-4); 16610 (1-1); 16611 (2-4); 16697 (3-6); 16771 (3-4); 16772 (6-2); 16829 (2-5); 17360 (7-2)

Hoffman, B., *et al.*, 730 (4-1)

Irwin, H.S., 269 (2-4); 289 (4-1); 621 (2-2); 643 (2-1); 758 (2-2); 769 (3-6)

Jansen-Jacobs, M.J. *et al.*, 129 (2-2); 313 (7-2); 588 (7-1); 590 (2-1); 636, 960 (6-2); 974 (7-2); 1178 (6-2); 1787 (7-2); 1817, 2506 (2-4); 2662 (2-1); 2768, 3805 (7-2); 4358 (2-2); 4851 (9-1)

Jenman, G.S., s.n. (3-2); 2151 (6-2); 3875 (1-1); 4789 (4-1); 4807 (3-6); 5063 (3-2); 5336 (2-4); 5337 (3-2); 5338 (3-6); 5339 (3-4); 5341, 5626 (4-1); 6247 (2-3); 7085 (6-2)

Kvist, L.P., *et al.*, 287 (7-2)

Lall, H. & M.S. Grewal, 347 (2-1)

Leechman, A., XVI (4-1)

Leng, H., 30, 58 (7-2); 420 (5-1)

Maas, P.J.M. *et al.*, 4119 (2-2); 7273 (2-1)

McDowell, T. *et al.*, 2396 (3-7); 4096 (7-2)

Omawale & R.N. Persaud, 79 (3-4); 96 (3-7)

Pipoly, J.J. *et al.*, 8348 (7-1); 11249 (3-1)

Robertson, K.R. & D.F. Austin, 317 (4-1); 325 (1-1)

Rothery, H.C., 62 (?) (4-1)

Schomburgk, Ro., s.n. (9-1); ser. I, 46 (10-2); ser. I, 586 (6-2); ser. I, 697 (7-1); ser. I, 702 (3-7); ser. I, 832 (9-1); ser. I, 934 (2-4); ser. II, 951 (2-3)

Schomburgk, Ri., **1527** (10-3)

Smith, A.C., 2393, 2504 (2-1)

Steege, H. ter, *et al.*, 307 (7-2)

Stockdale, F.A., 8827 (2-2)

Stoffers, A.L., A.R.A. Görts van Rijn *et al.*, 190 (7-2); 199, 247 (2-2); 338 (7-2); 497 (1-1)

Samuels, J.A., s.n., 48 (2-5); 98 (7-1); 340 (2-3)

Sastre, C. *et al.*, 8211 (10-1)

Sauvain, M., 389 (3-7)

Schulz, J.P., in LBB 11088 (2-5)

Soeprato, 2F, 32H (2-4); 37G (1-1); 46F (4-1); 180 (2-3); 197 (1-1); 216 (2-4); 370 (2-5); 371 (4-1); 379 (1-1); 397 (2-3)

Stahel, G., s.n. (2-2); s.n., s.n. (2-5); 24 (3-7); 88 (10-1)

Sterringa, J.T., in LBB 12390 (2-5); 12421 (4-1); 12448 (2-5)

Tawjoeran, J., in LBB 12346 (2-5)

Teunissen, P.A. *et al.*, in LBB 12611 (2-5); 14957 (1-1)

Tresling, J.H.A.T., 92 (7-1); 325 (2-4); 345 (3-7)

Tulleken, J.E., 20 (1-1); 302 (2-4); 584 (6-2)

UVS (Universiteit Suriname, see Kalpoe)

Versteeg, G.M., 271 (7-2); 288 (3-2); 543 (5-1)

Weigelt, C, s.n. (2-5)

Went, F.A.F.C., 171 (2-5); 299, 499 (3-6)

Wessels Boer, J.G., 419 (2-4); 495 (3-6); 497 (4-1); 563 (2-1); 566 (7-1)

Wullschlägel, H.R., 442 (3-6); 934 (2-3)

without coll., s.n. (2-5); s.n. (3-2); 4 (2-1); 4359 (8-1); 8562 (2-4)

FRENCH GUIANA

Alexandre, D.Y., 91 (3-6); 115 (2-1); 239 (2-5); 289, 290 (2-4); 315 (2-5); 380 (2-3); 490 (2-1)

Billiet, F.P. & B. Jadin, 4518 (2-1); 4745 (4-1)

Bordenave, B., 153 (4-1)

Broadway, W.E., 1 (2-5); 94 (4-1); 96 (2-5); 114 (3-4); 158 (2-1); 223 (3-6); 371 (2-3); 471, 568 (2-1); 570 (2-5); 688 (2-4)

Burgot, S-J., 1 (10-1)

Capus, F., 77 (7-2)

Cremers, G. *et al.*, 681 (4-1); 4744 (2-5); 7702 (4-1); 7780, 7786 (2-5); 7787 (2-4); 8429, 8430 (3-6); 8454 (3-4); 8457 (1-1); 8495, 8504 (2-5); 8561 (4-1); 8562, 8563 (2-5); 8678 (1-1); 8717 (2-4); 8740 (4-1); 9796 (2-1); 11343 (7-2); 12840 (3-7); 12855 (4-1); 12908 (2-5)

Dauchez, B., 13 (2-1)

Descoings, B. *et al.*, 20290 (2-5); 20461 (8-1)

Ducatillon, C. *et al.*, 60 (2-5); 78 (3-4)

Feuillet, C., 16 (2-1); 232 (2-5); 521 (7-2); 2142, 2175 (2-5)

Fleury, M., 266 (3-6); 684 (7-2); 692 (3-2); 720 (10-1)

Fournet, A., 100, 103 (2-1); 203 (3-4)

Garnier, F.A., 40 (7-2); 159 (2-1); 198 (10-1)

Gely, A., 38 (7-2); 74 (3-3)

Goff, A. le, 75 (2-5); 88, 180 (4-1); 186 (2-5)

Granville, J.J. de *et al.*, BC-64 (2-4); C-168 (7-2); 1956 (4-1); 1971 (3-1); 3177 (2-5); B-4957 (10-1); B-5067 (7-2); 5776 (3-5); 6621, 6644 (2-5)

Grenand, P., 149, 408 (7-2); 1619 (2-1); 1695 (2-5); 2144 (2-1)

Haxaire, C., 631 (7-2)

Hoff, M. *et al.*, 5074 (2-5); 5085 (3-6); 5107 (4-1); 5495 (2-5); 5497 (4-1); 5506 (2-1); 6084 (7-2); 6096 (2-4); 6214 (3-1); 6223 (2-5); 6281, 6282 (5-1); 6509 (7-2)

Jacquemin, H., 2216 (3-4); 2831 (10-1)
Kodjoed, J.F., 61 (10-1); 85 (2-1)
Leclerc, A., 15, 16, 17, 18, 19 (2-4); 46 (3-2); 47, 48 (3-7); 49 (3-2)
Leeuwenberg, A.J.M., 11696 (2-1)
Leng, H., 420 (5-1)
Lescure, J.P., 91, 306 (7-2)
Lindeman, J.C., 6778 (2-5)
Merlier, -, GY 202 (2-5); GY 222 (2-3); GY 246 (4-1)
Moretti, C., 108 (10-1)
Mori, S.A. *et al.*, 856 (7-2); 18261 (7-2); 18456 (2-4); 18658 (3-2); 18780 (2-5); 22074, 22346 (7-2); 23368 (2-4)
Oldeman, R.A.A., BC-25 (2-1); T-389 (3-2); B-443 (2-1); B-466 (7-2); B-735 (4-1); 2173 (3-6); B-2558, B-2677 (7-2); B-2875 (3-7); 3170 (4-1); B-3265 (7-2); B-3804 (4-1); B-3894 (10-1)
Prévost, M.F., 788 (7-2); 789 (2-4); 1197 (7-2)
Raynal-Roques, A., 19973 (7-2)
Rossignol, -, 4 (4-1)
Rothery, H.C., 61 (2-2)
Roubik, D., 114 (8-1); 128 (2-5)
Sagot, P.A., 481 (3-6); 482, 483 (3-7); 1321, 5506 (2-1)
Sastre, C. *et al.*, 119 (7-2); 8174 (10-1)
Sauvain, M., 64 (7-2)
Service Forestier, 4074 (2-4)
Skog, L. & C. Feuillet, 7291 (7-2); 7294 (2-4); 7315 (7-2); 7318 (2-2); 7542 (4-1)
Teunissen, P.A. *et al.* in LBB 12944 (2-1)
Toriola-Marbot, D. & M. Hoff, 249 (2-4)
Wittingthon, V., 101 (10-1)

without coll., 96 SA(?) (2-4);189, 215, 912 (2-1); AR.19724 (2-4); AR.19941 (4-1)

Portulacaceae

GUYANA

Appun, C.F., 981, 1763 (1-3)
Clarke, D., 2100 (2-1)
Cruz, J.S. de la, 966 (1-2); **996** (1-3); 2564 (1-2); **3693** (1-3); 3902 (1-2)
Gillespie, L.J., 1976 (1-5)
Graham, V., 371, 514 (1-3)
Harrison, S.G. & R. Persaud, 733, 878 (1-2)
Henkel, T.W., 3067 (1-5)
Hitchcock, A.S., 16590 (1-2); 16606 (1-4)
Jansen-Jacobs, M.J. *et al.*, 128 (1-3); 207, 665 (1-2); 667 (2-2); 3671 (1-5); 3886 (1-1); 3979, 4011 (1-5); 4021 (1-3); 4178 (1-2); 4758 (1-2); 4806 (1-5)
Jenman, G.S., 4508, 4540 (1-2); 4991, 5311 (1-4); 5312 (1-2)
Maas, P.J.M. *et al.*, 7635 (1-2)
McDowell, T., 2397 (1-2)
Mutchnik, P., 1182 (2-1); 1573 (2-2)
Omawale & R.N. Persaud, 71 (1-2)
without coll., s.n. (1-5)

SURINAME

Collector indigenus, 207 (1-2)
Dirven, J.G.P., LP 400 (1-2)
Doesburg, P.H. van, 49 (1-2)
Everaarts, A.P., 468 (1-2)
Florschütz, P.A. *et al.*, 969, 2662 (1-2)

Focke, H.C., 504, 704 (1-2)
Geijskes, D.C., 12 (1-2)
Hekking, W.H.A., 766 (1-2)
Herb. Schweinitz, s.n. (1-2)
Irwin, H.S. *et al.*, 55232, 57604 (1-2)
Jonker-Verhoef, A.M.E. & F.P. Jonker, 2, 659 (1-2)
Julen, C.R., in LBB 14557 (2-2)
Kalpoe, Ch., in UVS 16590 (1-2)
Kramer, K.U. & W.H.A. Hekking, 2374, 2644 (1-2)
Lanjouw, J., 277, 602, 1107 (1-2)
Lanjouw, J. & J.C. Lindeman, 3124 (1-2)
LBB (Lands Bosbeheer, see Julen and Sterringa)
Maguire, B., 23996 (1-5)
Oldenburger, F.H.F. *et al.*, 1342 (1-2)
Pulle, A., 71 (1-2)
Reijenga, Th.W., 485, 731 (1-2)
Samuels, J.A., 330 (1-2)
Sauvain, M., 609, 706 (1-2)
Soeprato, 37 (1-2)
Stahel, G. & J.W. Gonggrijp, in BW 2947 (1-2)
Sterringa, J.T., in LBB 12475 (1-2)
Tresling, J.H.A.T., 322 (1-2)
UVS (Universiteit Suriname, see Kalpoe)
Vaart, J. van der, 1519 (1-2)
Versteeg, G.M., 922 (1-2)
Wessels Boer, J.G., 529, 1157 (1-2)

FRENCH GUIANA

Alexandre, D.Y., 352 (1-2)
Billiet, F.P. & B. Jadin, 4738 (2-2)
Bordenave, B., 165 (2-2)
Broadway, W.E., 100 (1-2); 105 (1-3); 835 (1-5); 962 (1-3)
Burgot, S-J., 8 (1-2)

Cremers, G.A. *et al.*, 4642 (1-5); 5040 (2-2); 7740 (1-3); 7754 (2-2); 8415, 8416 (1-5); 8433 (1-2); 8451 (1-3); 8499, 8513 (2-2); 11844 (1-5)
Fleury, M., 133 (1-2)
Goff, A. le, 74 (1-5); 96 (2-2)
Granville, J.J. de *et al.*, 1357, B-2563, 4339 (1-5); 5362, 5364, 5754 (2-2); 7261 (1-3); 9781, 9800, 9874 (1-5); 50004 (1-3)
Grenand, P., 1370, 1470 (1-2)
Hoff, M., 5091 (2-1); 5106 (1-1); 5709 (1-2)
Jacquemin, H., 2130 (2-2)
Kodjoed, J.F., 96 (1-2)
Leclerc, A., 54 (1-2)
Lindeman, J.C., 6762 (1-2)
Oldeman, R.A.A., B-734 (1-2)
Prévost, M.F., 1325 (1-1); 2176 (2-2)
Raynal-Roques, A., 20042, 21518 (1-5)
Rothery, H.C., 111 (1-3)
Sastre, C. *et al.*, 4208 (1-3)
Sauvain, M., 623 (1-2)
Tostain, O., 76 (1-2); 77 (1-5)
without coll., 49 (1-2); 1290 (1-2); 4738 (2-2)

Basellaceae

GUYANA

Hitchcock, A.S., 17401 (2-1)
Omawale & R.N. Persaud, 56, 57, 59 (2-1)
Parker, C.S., s.n. (2-1)

SURINAME

Hekking, W.H.A., 1132 (2-1)

Kramer, K.U. & W.H.A. Hekking, 2861 (2-1)

FRENCH GUIANA

Jacquemin, H., 2602 (1-1)
Prévost, M.F., 1482 (2-1)
Wittingthon, V., 89 (1-1)

Molluginaceae

GUYANA

Davis, D.H., 101 (2-1)
Goodland, R.J.A., 1054 (1-1)
Harrison, S.G. *et al.*, 1071, 1765 (2-1)
Irwin, H.S., 773 (1-1)
Jansen-Jacobs, M.J. *et al.*, 125 (1-1)
Kortright, P., 135 (2-1)
Parker, C.S., s.n. (2-1)
Pulle, A., II 543 (2-1)
Rothery, H.C., 109 (2-1)
Schomburgk, Ro., ser. I, 255 (2-1)
Schulz, J.P., 8309 (2-1)
Thurn, E.F. im, s.n. (2-1)
without coll., s.n. (2-1)

SURINAME

Boerboom, J., in LBB 9522 (2-1)
BW (Boswezen Suriname, see Stahel & Gonggrijp)
Donselaar, J. van & W.A.E. van Donselaar, 379 (2-1)
Focke, H.C., 274 (2-1)
Geijskes, D.C., 193 (2-1)
Hekking, W.H.A., 1090 (2-1)
Heyligers, P.C., 742 (2-1)
Hostmann, F.W.R., 1019, 5101 (2-1)
Jonker-Verhoef, A.M.E. & F.P. Jonker, 158 (2-1)

Lanjouw, J. & J.C. Lindeman, 999, 1818, 1847 (2-1)
LBB (Lands Bosbeheer, see Boerboom, Pons and Schulz)
Mennega, A.M.W., 402 (2-1)
Pons, T.L., in LBB 12654 (2-1)
Schulz, J.P., in LBB 9930 (2-1)
Stahel, G. & J.W. Gonggrijp, in BW 3035 (2-1)
Versteeg, G.M., 272 (2-1)
Went, F.A.F.C., 418, 465 (2-1)

FRENCH GUIANA

Broadway, W.E., 507 (2-1)
Cremers, G.A., 635, 4922, 5120, 7996, 11439, 12463, 12896 (2-1)
Feuillet, C., 579 (2-1)
Granville, J.J. de, 1957 (2-1)
Hoff, M., 5244 (2-1)
Jelski, C. von, s.n. (2-1)
Leprieur, F.R., 7038 (2-1)
Lescure, J.P., 635 (2-1)
Merlier, -, GY 128, GY 130 (2-1)
Oldeman, R.A.A., 171h (1-1)
Philippe, M., 91 (2-1)
Poiteau, P.A., s.n. (2-1)
Sagot, P.A., 37 (2-1)
Sastre, C. *et al.*, 234 (2-1)

Caryophyllaceae

GUYANA

Appun, C.F., 841 (1-1), 2121 (2-1)
Cooper, A., 310 (2-1)
Fanshawe, D.B., 64 (2-1)
Georgetown Botanic Garden, H 1759/51 (1-1)
Goodland, R., 537 (2-1)
Hahn, W. *et al.*, 4472 (1-1)
Harrison, S.G. & R. Persaud, 751, 1026 (2-1)

Hohenkerk, L.S., 127, 964 (2-1)
Irwin, H.S., 342 (2-1)
Jenman, G.S., 483 (2-1)
Maas, P.J.M. *et al.*, 3770 (2-1)
Maguire, B. & D.B. Fanshawe, 23562 (1-1)
Mell, R.C. & C.D. Mell, 232 (1-1)
Pipoly, J.J. *et al.*, 9376 (2-1)
Pulle, A., II 533 (2-1)

SURINAME

Archer, W.A., 2837 (2-1)
Caulon, B., 538 (1-1)
Doesburg, P.H. van, 27 (1-1)
Donselaar, J. van & W.A.E. van Donselaar, 331 (2-1)
Essed, E, s.n., 81 (1-1)
Heinemann, D., s.n. (1-1)
Herb. Schweinitz, s.n. (1-1)
Hostmann, F.W.R. *et al.*, 303 (1-1), **598** (2-1), 714, 714a (1-1)
Jansma, R., 9 (2-1)
Kappler, A., 25 (2-1)
Lanjouw, J. & J.C. Lindeman, 1663, 1708 (2-1)
LBB (Lands Bosbeheer, see Teunissen)
Lindeman, J.C. & E.A. Mennega, 182 (2-1); 5722 (1-1)
Maas, P.J.M. *et al.*, 7693 (2-1)
Reijenga, Th.W., 624 (1-1)
Teunissen, P.A. *et al.*, in LBB 14521 (2-1)
without coll., s.n. (1-1)

FRENCH GUIANA

Alexandre, D.Y., 383 (1-1)
Broadway, W.E., 27 (1-1)
Ducatillon, C. & A. Gelly, 83 (1-1)
Feuillet, C., 1696 (1-1)

Jacquemin, H., 1497, 1509, 1636 (1-1)
Moretti, C., 268 (1-1)
Mori, S.A. *et al.*, 23381 (1-1)
Poiteau, P.A., s.n. (1-1)
Raynal-Roques, A., 19874 (1-1)

Sarraceniaceae

GUYANA

Abbensetts, N.J., 3, 42 (1-1)
Hahn, W.J. *et al.*, 5528 (1-1)
Renz, J., 14202, 14267 (1-1)
Schomburgk, Ri., 983 (1-1)
Tate, G.H.H., 363 (1-1)
Thurn, E.F. im, 258 (1-1)

EXTRA GUIANAN

Schomburgk, Ro., **ser. I, 1050** (1-1) – VEN
Ule, E., Herbarium Brasiliense 8608 (1-1) – BRA

Droseraceae

GUYANA

Abbensetts, N.J., 11 (1-7)
Appun, C.F., 1157 (1-7)
Cook, C.D.K., 87, 179 (1-8)
Cruz, J.S. de la, 4012 (1-5)
Hahn, W. *et al.*, 4468 (1-6); 5509 (1-7)
Jansen-Jacobs, M.J. *et al.*, 455 (1-8); 1392 (1-2); 2743, 2744 (1-8); 3680 (1-2); 3157 (1-4); 5023, 5040 (1-8)
Jenman, G.S., 912, 1293 (1-1); 2255 (1-5); 3769 (1-2)

Liesner, R., 23237 (1-7)
Maas, P.J.M. *et al.*, 3682, 4022 (1-8); 4268, 4346 (1-6); 4398 (1-5); 5660 (1-7); 5529, 5659, 5740, 7619 (1-5); 7717 (1-8)
Maguire, B. *et al.*, **23466**, 32286, 32517 (1-6); 40645 (1-7); 45979-A, 45980-A (1-6)
McConnell, F.V. & J.J. Quelch, 154, 681 (1-7)
Pipoly, J.J. *et al.*, 11133 (1-7)
Schomburgk, Ro., **ser. I, 102** (1-8)
Smith, A.C., 2301 (1-8)
Thurn, E.F. im, 313 (1-7)
Tillett, S.S. *et al.*, 44887 (1-6)
Ule, E.H.G., 8610 (1-7)
Wilson-Browne, G., 279 (1-8)
without col., (1-5)
without col., (1-7)

SURINAME

Donselaar, J. van, 406 (1-2); 739 (1-3); 2938 (1-2); 2973 (1-3); 3664 (1-8)
Geijskes, D.C., s. n. (1-3)
Heyligers, P.C., 365 (1-2)
Florschütz, P.A. *et al.*, 247, 622, 799, 1984, 1986 (1-2)

Jansma, R., 25, 37 (1-2)
Kramer, K.U. & W.H.A. Hekking, 2938 (1-2), 3041, 3260, 3280, (1-3)
Kuyper, J., 16 (1-2)
Lanjouw, J., 146 (1-2)
Lanjouw, J. & J.C. Lindeman, 587, 686, 855 (1-2); 852, 903, 944 (1-5)
Maas, P.J.M., 3202, 3310 (1-2)
Maguire, B. *et al.*, 24432, 24486 (1-5); 25035 (1-2)
Teunissen, P.A. *et al.*, in LBB 11495, 12169 (1-2)
without col., (1-2)

FRENCH GUIANA

Cremers, G. *et al.*, 9544, 10683 (1-2)
Donselaar, J. van, 2595 (1-2)
Gentry, A. & E. Zardini, 50330 (1-2)
Leprieur, F.R., 145 (1-3)
Sastre, C., 11 (1-3); 1322 (1-2)

INDEX TO SYNONYMS, NAMES IN NOTES AND SOME TYPES

Phytolaccaceae

ACHATOCARPACEAE, see family description, note
Achatocarpus
    pubescens C.H. Wright, see family description, note
Ancistrocarpus
    maypurensis Kunth = 1-2
CHENOPODIACEAE, see 1, note
Phytolacca
    americana L., see 3-2, note; see 3, type
    decandra L., see 3-2, note
Rivina
    octandra L. = 6-1; see 6, type
Seguieria
    aculeata Jacq., see 5-1, note
    foliosa Benth. = 5-1
Trichostigma
    rivinoides A. Rich., see 6, type

Nyctaginaceae

Boerhavia
    caribaea Jacq., see 1-1, note
    coccinea Mill., see 1-1, note
    decumbens Vahl, see 1-1, note
    erecta L., lectotype 1
    hirsuta Willd., see 1-1, note
    paniculata Rich. = 1-1; see 1-1, note
    surinamensis Miq. = 1-1
    viscosa Lag. & Rodr., see 1-1, note
Bougainvillea
    spectabilis Willd. var. *glabra* (Choisy) Hook. = 2-1
    x buttiana Holttum & Standley, see 2-2, note
Guapira
    amacurensis Steyerm., see 3-3, note
    glabra (Heimerl) Steyerm. = 3-2
    graciliflora (Mart.) Lundell, see 3-4, note
    guianensis Aubl., type 3; see 3-2, note
    heimerliana (Standl.) Lundell = 3-4
    marcano-bertii Steyerm., see 3-3, note
    olfersiana (Link, Klotzsch & Otto) Lundell, see 3-2 and 3-4, note

Aizoaceae

Portulaca
  portulacastrum L. = 1-1
Sesuvium
  acutifolium Miq. = 1-1

Chenopodiaceae

Chenopodium
  rubrum L., see 1, type

Amaranthaceae

Achyranthes
  altissima Jacq. = 6-2
  aspera L. var. *indica* L. = 1-1
  aspera L. var. pubescens (Moq.) C.C. Towns., see 1-1, note
  halimifolia Lam. = 2-3
  indica (L.) Mill. = 1-1
  maritima (Mart.) Standley, see 2-4, note
  prostrata L. = 7-2
Acnida L. = 3
  australis A. Gray = 3-1
  cuspidata Bertero ex Spreng. = 3-1
Alternanthera
  achyranthes Forssk., see 2, type
  bettzickiana (Regel) G. Nicholson = 2-2
  'Bettzickiana', see 2-2, note
  crucis (Moq.) Bold. = 2-3
  dentata R.E. Fr. = 2-1
  dentata 'Rubiginosa', see 2-1, note
  dentata 'Ruby', see 2-1, note
  ficoidea (L.) P. Beauv. var. *amoena* (Lem.) L.B. Sm. & Downs = 2-2
  ficoidea 'Amoena', see 2-2, note
  ficoidea (L.) P. Beauv. var. *bettzickiana* (Regel) Backer = 2-2
  flavogrisea (Urb.) Urb. = 2-3
  littoralis P. Beauv. var. *maritima* (Mart.) Pedersen, see 2-4, note
  maritima (Mart.) A. St.-Hil., see 2-4, note
  paronychioides A. St.-Hil., see 2-2, note
  ramosissima (Mart.) Chodat & Hassl., see 2-1, note
  sessilis (L.) DC. var. *amoena* Lem. = 2-2
  tenella Colla = 2-2
  tenella Colla subsp. *flavogrisea* (Urb.) Mears & Veldkamp = 2-3

Amaranthus
gracilis Desf. = 3-7
lividus L. = 3-2
melancholicus L., see 3-7, note
paniculatus L. = 3-5
polygamus L., see 3-7, note
tricolor L., see 3-7, note
Blutaparon
repens Raf., see 4, type
vermiculare (L.) Mears var. aggregatum (Willd.) Mears, see 4-1, note
vermiculare (L.) Mears var. longispicatum (Moq.) Mears, see 4-1, note
Bucholzia
philoxeroides Mart. = 2-4
Celosia
argentea L. f. *cristata* (L.) Schinz = 5-1
argentea L. var. *cristata* (L.) O. Kuntze = 5-1
argentea 'Plumosa', see 5-1, note
cristata L. = 5-1
'Cristata', see 5-1, note
Chenopodium
caudatum Jacq. = 3-7
Desmochaeta
achyranthoides Kunth = 7-1
Gomphrena
brasiliana L. = 2-1
dentata Moench = 2-1
ficoidea L. = 2-2
sessilis L. = 2-5
stenophylla Spreng. = 10-2
vermicularis L. = 4-1
Iresine
glomerata Spreng. = 10-2
grandiflora Hook. = 10-3
polymorpha Mart. = 9-1
surinamensis Moq. = 4-1; see 4-1, note
vermicularis (L.) Moq. = 4-1
Pfaffia
glauca (Mart.) Spreng. = 10-2
grandiflora (Hook.) R.E. Fr. var. *hookeriana* (Hemsl.) O. Stützer f.
    guianensis Klotzsch ex O. Stützer = 10-3
grandiflora (Hook.) R.E. Fr. var. *typica* O. Stützer = 10-3
stenophylla (Spreng.) Stuchlik var. *foliosa* O. Stützer = 10-2
stenophylla (Spreng.) Stuchlik = 10-2

Philoxerus
  vermicularis (L.) Sm. = 4-1
Serturnera
  glauca Mart. = 10-2
Telanthera R. Br. = 2
  bettzickiana Regel = 2-2
  crucis Moq. = 2-3
  dentata Moq. = 2-1
  flavogrisea Urb. = 2-3
  manillensis Walp., see 2, type
  maritima (Mart.) Moq., see 2-4, note

Portulacaceae

Portulaca
  cayennensis D. Legrand = 1-5
  fruticosa L. = 2-1
  grandiflora Hook. var. immersostellulata (Poelln.) D. Legrand, see 1-1, note
  grandiflora Hook. var. grandiflora f. depressa (D. Legrand) D. Legrand, see 1-1, note
  lanata Rich. = 1-3
  mucronata Link, see 1-2, note
  oleracea L. var. *granulatostellata* Poelln. = 1-2
  paniculata Jacq. = 2-2
  pilosa L. var. *guyannensis* D. Legrand = 1-3
  pilosa L. subsp. *grandiflora* (Hook.) R. Geesink = 1-1
  rubricaulis Kunth, see 1-3, note
  triangularis Jacq. = 2-1
  teretifolia Kunth, see 1-3, note
Talinum
  triangulare (Jacq.) Willd. = 2-1; see 2, type

Basellaceae

Anredera
  vesicaria (Lam.) C.F. Gaertn., see 1, type
  spicata J.F. Gmel., see 1, type
Basella
  rubra L., see 2, type
Boussingaultia
  leptostachys Moq. = 1-1

Molluginaceae

AIZOACEAE, see family description, note
Glinus
  lotoides L., see 1, type
Mollugo
  radiata Ruiz & Pav. = 1-1

Caryophyllaceae

Achyranthes
  corymbosa (L.) Lam. = 2-1
Drymaria
  arenarioides Humb. & Bonpl. ex Schult., see 1, type
Holosteum
  cordatum L. = 1-1
Polycarpaea
  atherophora Steud. = 2-1
  corymbosa (L.) Lam. var. brasiliensis (Cambess.) Chodat & Hassl.,
    see 2-1, note
  teneriffae Lam., see 2, type

Sarraceniaceae

Heliamphora
  minor Gleason, see 1-1, note
  macdonaldae Gleason, see 1, note
  tyleri Gleason see 1, note

Droseraceae

Drosera
  colombiana A. Fernández = 1-3
  dentata Benth. = 1-8
  meristocaulis Maguire & Wurdack, see 1, subdivision
  montana A. St.-Hil. var. *robusta* Diels = 1-7
  montana A. St.-Hil. var. *roraimae* Diels = 1-7
  panamensis M.D. Correa & A.S. Taylor = 1-3
  pusilla Kunth = 1-1
  sanariapoana Steyerm. = 1-3
  tenella Humb. & Bonpl. ex Roem. & Schult. = 1-2
  tenella Humb. & Bonpl. ex Roem. & Schult. var. *esmeraldae*
    Steyerm. = 1-4

INDEX TO VERNACULAR NAMES

glycerine 1-1
green stem poi 2-1
malabar spinach 2-1
poi 2-1
poisang 2-1
purple stem poi 2-1
spinazie 2-1

Molluginaceae

sulumonti 2-1

Caryophyllaceae

mignonette 1-1
petit quinine 1-1
piki fowroesopo 1-1
timignonette 1-1

Droseraceae

sundew 1
yeberu (Arow.) 1-2
yeberubina (Arow.) 1-2

Alphabetic list of families of series A occurring in the Guianas

Defined as in Cronquist, 1981, and numbered in his sequence, with alternative names. Those published, with chronological fascicle number and year.

Abolbodaceae			Campanulaceae	162	
(see Xyridaceae	182)	15. 1994	(incl. Lobeliaceae)		
Acanthaceae	156		Cannaceae	195	1. 1985
(incl. Thunbergiaceae)			Canellaceae	004	
(excl. Mendonciaceae	159)		Capparaceae	067	
Achatocarpaceae	028	22. 2003	Caprifoliaceae	164	
Agavaceae	202		Caricaceae	063	
Aizoaceae	030	22. 2003	Caryocaraceae	042	
(excl. Molluginaceae	036)	22. 2003	Caryophyllaceae	037	22. 2003
Alismataceae	168		Casuarinaceae	026	11. 1992
Amaranthaceae	033	22. 2003	Cecropiaceae	022	11. 1992
Amaryllidaceae			Celastraceae	109	
(see Liliaceae	199)		Ceratophyllaceae	014	
Anacardiaceae	129	19. 1997	Chenopodiaceae	032	22. 2003
Anisophylleaceae	082		Chloranthaceae	008	
Annonaceae	002		Chrysobalanaceae	085	2. 1986
Apiaceae	137		Clethraceae	072	
Apocynaceae	140		Clusiaceae	047	
Aquifoliaceae	111		(incl. Hypericaceae)		
Araceae	178		Cochlospermaceae		
Araliaceae	136		(see Bixaceae	059)	
Arecaceae	175		Combretaceae	100	
Aristolochiaceae	010	20. 1998	Commelinaceae	180	
Asclepiadaceae	141		Compositae		
Asteraceae	166		(= Asteraceae	166)	
Avicenniaceae			Connaraceae	081	
(see Verbenaceae	148)	4. 1988	Convolvulaceae	143	
Balanophoraceae	107	14. 1993	(excl. Cuscutaceae	144)	
Basellaceae	035	22. 2003	Costaceae	194	1. 1985
Bataceae	070		Crassulaceae	083	
Begoniaceae	065		Cruciferae		
Berberidaceae	016		(= Brassicaceae	068)	
Bignoniaceae	158		Cucurbitaceae	064	
Bixaceae	059		Cunoniaceae	081a	
(incl. Cochlospermaceae)			Cuscutaceae	144	
Bombacaceae	051		Cycadaceae	208	9. 1991
Bonnetiaceae			Cyclanthaceae	176	
(see Theaceae	043)		Cyperaceae	186	
Boraginaceae	147		Cyrillaceae	071	
Brassicaceae	068		Dichapetalaceae	113	
Bromeliaceae	189	p.p. 3. 1987	Dilleniaceae	040	
Burmanniaceae	206	6. 1989	Dioscoreaceae	205	
Burseraceae	128		Dipterocarpaceae	041a	17. 1995
Butomaceae			Droseraceae	055	22. 2003
(see Limnocharitaceae	167)		Ebenaceae	075	
Byttneriaceae			Elaeocarpaceae	048	
(see Sterculiaceae	050)		Elatinaceae	046	
Cabombaceae	013		Eremolepidaceae	105a	
Cactaceae	031	18. 1997	Ericaceae	073	
Caesalpiniaceae	088	p.p. 7. 1989	Eriocaulaceae	184	
Callitrichaceae	150		Erythroxylaceae	118	

Euphorbiaceae	115		Loganiaceae	138	
Euphroniaceae	123a	21. 1998	Loranthaceae	105	
Fabaceae	089		(excl. Viscaceae	106)	
Flacourtiaceae	056		Lythraceae	094	
(excl. Lacistemaceae	057)		Malpighiaceae	122	
(excl. Peridiscaceae	058)		Malvaceae	052	
Gentianaceae	139		Marantaceae	196	
Gesneriaceae	155		Marcgraviaceae	044	
Gnetaceae	209	9. 1991	Martyniaceae		
Gramineae			(see Pedaliaceae	157)	
(= Poaceae	187)	8. 1990	Mayacaceae	183	
Gunneraceae	093		Melastomataceae	099	13. 1993
Guttiferae			Meliaceae	131	
(= Clusiaceae	047)		Mendonciaceae	159	
Haemodoraceae	198	15. 1994	Menispermaceae	017	
Haloragaceae	092		Menyanthaceae	145	
Heliconiaceae	191	1. 1985	Mimosaceae	087	
Henriquesiaceae			Molluginaceae	036	22. 2003
(see Rubiaceae	163)		Monimiaceae	005	
Hernandiaceae	007		Moraceae	021	11. 1992
Hippocrateaceae	110	16. 1994	Moringaceae	069	
Humiriaceae	119		Musaceae	192	1. 1985
Hydrocharitaceae	169		(excl. Strelitziaceae	190)	
Hydrophyllaceae	146		(excl. Heliconiaceae	191)	
Icacinaceae	112	16. 1994	Myoporaceae	154	
Hypericaceae			Myricaceae	025	
(see Clusiaceae	047)		Myristicaceae	003	
Iridaceae	200		Myrsinaceae	080	
Ixonanthaceae	120		Myrtaceae	096	
Juglandaceae	024		Najadaceae	173	
Juncaginaceae	170		Nelumbonaceae	011	
Krameriaceae	126	21. 1998	Nyctaginaceae	029	22. 2003
Labiatae			Nymphaeaceae	012	
(= Lamiaceae	149)		(excl. Nelumbonaceae	010)	
Lacistemaceae	057		(excl. Cabombaceae	013)	
Lamiaceae	149		Ochnaceae	041	
Lauraceae	006		Olacaceae	102	14. 1993
Lecythidaceae	053	12. 1993	Oleaceae	152	
Leguminosae			Onagraceae	098	10. 1991
(= Mimosaceae	087)		Opiliaceae	103	14. 1993
+ Caesalpiniaceae	088)	p.p. 7. 1989	Orchidaceae	207	
+ Fabaceae	089)		Oxalidaceae	134	
Lemnaceae	179		Palmae		
Lentibulariaceae	160		(= Arecaceae	175)	
Lepidobotryaceae	134a		Pandanaceae	177	
Liliaceae	199		Papaveraceae	019	
(incl. Amaryllidaceae)			Papilionaceae		
(excl. Agavaceae	202)		(= Fabaceae	089)	
(excl. Smilacaceae	204)		Passifloraceae	062	
Limnocharitaceae	167		Pedaliaceae	157	
(incl. Butomaceae)			(incl. Martyniaceae)		
Linaceae	121		Peridiscaceae	058	
Lissocarpaceae	077		Phytolaccaceae	027	22. 2003
Loasaceae	066		Pinaceae	210	9. 1991
Lobeliaceae			Piperaceae	009	
(see Campanulaceae	162)		Plantaginaceae	151